Tianran Wang
Yi Zhang
Haibin Yu
Feiyue Wang

# Advanced Manufacturing Technology in China: A Roadmap to 2050

Chinese Academy of Sciences

Tianran Wang

Yi Zhang

Haibin Yu

Feiyue Wang

*Editors*

# Advanced Manufacturing Technology in China: A Roadmap to 2050

With 13 figures

Science Press
Beijing

Springer

*Editors*

Tianran Wang
Shenyang Institute of Automation, CAS
110016, Shenyang, China
E-mail: trwang@sia.cn

Haibin Yu
Shenyang Institute of Automation, CAS
110016, Shenyang, China
E-mail: yhb@sia.cn

Yi Zhang
Institute of Process Engineering, CAS
100190, Beijing, China
E-mail: yizhang@home.ipe.ac.cn

Feiyue Wang
Institute of Automation, CAS
100190, Beijing, China
E-mail: feiyuewang@gmail.com

ISBN 978-7-03-027541-7
Science Press Beijing

ISBN 978-3-642-13854-6          e-ISBN 978-3-642-13855-3
Springer Heidelberg Dordrecht London New York

Library of Congress Control Number: 2010928339

# Members of the Editorial Committee and the Editorial Office

## Editor-in-Chief

Yongxiang Lu

## Editorial Committee

Yongxiang Lu   Chunli Bai   Erwei Shi   Xin Fang   Zhigang Li   Xiaoye Cao   Jiaofeng Pan

# Research Group on Advanced Manufacturing Technology of the Chinese Academy of Sciences

**Chairman:** Tianran Wang   Shenyang Institute of Automation, CAS

**Vice Chairman:** Yi Zhang   Institute of Process Engineering, CAS

**Members:** ( In the alphabetical order of the Chinese surname )

| | |
|---|---|
| Dong Yu | Shenyang Institute of Computing Technology, CAS |
| Haibin Yu | Shenyang Institute of Automation, CAS |
| Tingcan Ma | The Wuhan Branch of the National Science Library, CAS |
| Feiyue Wang | Institute of Automation, CAS |
| Yunlong Zhu | Shenyang Institute of Automation, CAS |
| Huiquan Li | Institute of Process Engineering, CAS |
| Chao Yang | Institute of Process Engineering, CAS |
| Yanwu Yang | Institute of Automation, CAS |
| Tao Ku | Shenyang Institute of Automation, CAS |
| Hong Song | Shenyang Institute of Automation, CAS |
| Hongzhang Chen | Institute of Process Engineering, CAS |
| Wensheng Zhang | Institute of Automation, CAS |
| Xiangping Zhang | Institute of Process Engineering, CAS |
| Shili Zheng | Institute of Process Engineering, CAS |
| Fen Xia | Institute of Automation, CAS |
| Hongbin Cao | Institute of Process Engineering, CAS |
| Xiaowei Peng | Institute of Process Engineering, CAS |
| Peng Zeng | Shenyang Institute of Automation, CAS |
| Gang Xiong | Institute of Automation, CAS |

# Foreword to the Roadmaps 2050[*]

China's modernization is viewed as a transformative revolution in the human history of modernization. As such, the Chinese Academy of Sciences (CAS) decided to give higher priority to the research on the science and technology (S&T) roadmap for priority areas in China's modernization process. What is the purpose? And why is it? Is it a must? I think those are substantial and significant questions to start things forward.

## Significance of the Research on China's S&T Roadmap to 2050

We are aware that the National Mid- and Long-term S&T Plan to 2020 has already been formed after two years' hard work by a panel of over 2000 experts and scholars brought together from all over China, chaired by Premier Wen Jiabao. This clearly shows that China has already had its S&T blueprint to 2020. Then, why did CAS conduct this research on China's S&T roadmap to 2050?

In the summer of 2007 when CAS was working out its future strategic priorities for S&T development, it realized that some issues, such as energy, must be addressed with a long-term view. As a matter of fact, some strategic researches have been conducted, over the last 15 years, on energy, but mainly on how to best use of coal, how to best exploit both domestic and international oil and gas resources, and how to develop nuclear energy in a discreet way. Renewable energy was, of course, included but only as a supplementary energy. It was not yet thought as a supporting leg for future energy development. However, greenhouse gas emissions are becoming a major world concern over

---

* It is adapted from a speech by President Yongxiang Lu at the first High-level Workshop on China's S&T Roadmap for Priority Areas to 2050, organized by the Chinese Academy of Sciences, in October, 2007.

the years, and how to address the global climate change has been on the agenda. In fact, what is really behind is the concern for energy structure, which makes us realize that fossil energy must be used cleanly and efficiently in order to reduce its impact on the environment. However, fossil energy is, pessimistically speaking, expected to be used up within about 100 years, or optimistically speaking, within about 200 years. Oil and gas resources may be among the first to be exhausted, and then coal resources follow. When this happens, human beings will have to refer to renewable energy as its major energy, while nuclear energy as a supplementary one. Under this situation, governments of the world are taking preparatory efforts in this regard, with Europe taking the lead and the USA shifting to take a more positive attitude, as evidenced in that: while fossil energy has been taken the best use of, renewable energy has been greatly developed, and the R&D of advanced nuclear energy has been reinforced with the objective of being eventually transformed into renewable energy. The process may last 50 to 100 years or so. Hence, many S&T problems may come around. In the field of basic research, for example, research will be conducted by physicists, chemists and biologists on the new generation of photovoltaic cell, dye-sensitized solar cells (DSC), high-efficient photochemical catalysis and storage, and efficient photosynthetic species, or high-efficient photosynthetic species produced by gene engineering which are free from land and water demands compared with food and oil crops, and can be grown on hillside, saline lands and semi-arid places, producing the energy that fits humanity. In the meantime, although the existing energy system is comparatively stable, future energy structure is likely to change into an unstable system. Presumably, dispersive energy system as well as higher-efficient direct current transmission and storage technology will be developed, so will be the safe and reliable control of network, and the capture, storage, transfer and use of $CO_2$, all of which involve S&T problems in almost all scientific disciplines. Therefore, it is natural that energy problems may bring out both basic and applied research, and may eventually lead to comprehensive structural changes. And this may last for 50 to 100 years or so. Taking the nuclear energy as an example, it usually takes about 20 years or more from its initial plan to key technology breakthroughs, so does the subsequent massive application and commercialization. If we lose the opportunity to make foresighted arrangements, we will be lagging far behind in the future. France has already worked out the roadmap to 2040 and 2050 respectively for the development of the 3rd and 4th generation of nuclear fission reactors, while China has not yet taken any serious actions. Under this circumstance, it is now time for CAS to take the issue seriously, for the sake of national interests, and to start conducting a foresighted research in this regard.

This strategic research covers over some dozens of areas with a long-term view. Taking agriculture as an example, our concern used to be limited only to the increased production of high-quality food grains and agricultural by-products. However, in the future, the main concern will definitely be given to the water-saving and ecological agriculture. As China is vast in territory, diversified technologies in this regard are the appropriate solutions. Animal husbandry has been used by developed countries, such as Japan and Denmark, to make bioreactor and pesticide as well. Plants have been used by Japan to make bioreactors which are safer and cost-effective than that made from animals. Potato, strawberry, tomato and the like have been bred in germ-free greenhouses, and value-added products have been made through gene transplantation technology. Agriculture in China must not only address the food demands from its one billions-plus population, but also take into consideration of the value-added agriculture by-products and the high-tech development of agriculture as well. Agriculture in the future is expected to bring out some energies and fuels needed by both industry and man's livelihood as well. Some developed countries have taken an earlier start to conduct foresighted research in this regard, while we have not yet taken sufficient consideration.

Population is another problem. It will be most likely that China's population will not drop to about 1 billion until the end of this century, given that the past mistakes of China's population policy be rectified. But the subsequent problem of ageing could only be sorted out until the next century. The current population and health policies face many challenges, such as, how to ensure that the 1.3 to 1.5 billion people enjoy fair and basic public healthcare; the necessity to develop advanced and public healthcare and treatment technologies; and the change of research priority to chronic diseases from infectious diseases, as developed countries have already started research in this regard under the increasing social and environmental change. There are many such research problems yet to be sorted out by starting from the basic research, and subsequent policies within the next 50 years are in need to be worked out.

Space and oceans provide humanity with important resources for future development. In terms of space research, the well-known Manned Spacecraft Program and China's Lunar Exploration Program will last for 20 or 25 years. But what will be the whole plan for China's space technology? What is the objective? Will it just follow the suit of developed countries? It is worth doing serious study in this regard. The present spacecraft is mainly sent into space with chemical fuel propellant rocket. Will this traditional propellant still be used in future deep space exploration? Or other new technologies such as electrical propellant, nuclear energy propellant, and solar sail technologies be developed? We haven't yet done any strategic research over these issues, not even worked out any plans. The ocean is abundant in mineral resources, oil and gas, natural gas hydrate, biological resources, energy and photo-free biological evolution, which may arise our scientific interests. At present, many countries have worked out new strategic marine plans. Russia, Canada, the USA, Sweden and Norway have centered their contention upon the North Pole, an area of strategic significance. For this, however, we have only limited plans.

The national and public security develops with time, and covers both conventional and non-conventional security. Conventional security threats only refer to foreign invasion and warfare, while, the present security threat may come out from any of the natural, man-made, external, interior, ecological, environmental, and the emerging networking (including both real and virtual) factors. The conflicts out of these must be analyzed from the perspective of human civilization, and be sorted out in a scientific manner. Efforts must be made to root out the cause of the threats, while human life must be treasured at any time.

In general, it is necessary to conduct this strategic research in view of the future development of China and mankind as well. The past 250 years' industrialization has resulted in the modernization and better-off life of less than 1 billion people, predominantly in Europe, North America, Japan and Singapore. The next 50 years' modernization drive will definitely lead to a better-off life for 2–3 billion people, including over 1 billion Chinese, doubling or tripling the economic increase over that of the past 250 years, which will, on the one hand, bring vigor and vitality to the world, and, on the other hand, inevitably challenge the limited resources and eco-environment on the earth. New development mode must be shaped so that everyone on the earth will be able to enjoy fairly the achievements of modern civilization. Achieving this requires us, in the process of China's modernization, to have a foresighted overview on the future development of world science and human civilization, and on how science and technology could serve the modernization drive. S&T roadmap for priority areas to 2050 must be worked out, and solutions to core science problems and key technology problems must be straightened out, which will eventually provide consultations for the nation's S&T decision-making.

## Possibility of Working out China's S&T Roadmap to 2050

Some people held the view that science is hard to be predicted as it happens unexpectedly and mainly comes out of scientists' innovative thinking, while, technology might be predicted but at the maximum of 15 years. In my view, however, S&T foresight in some areas seems feasible. For instance, with the exhaustion of fossil energy, some smart people may think of transforming solar energy into energy-intensive biomass through improved high-efficient solar thin-film materials and devices, or even developing new substitute. As is driven by huge demands, many investments will go to this emerging area. It is, therefore, able to predict that, in the next 50 years, some breakthroughs will undoubtedly be made in the areas of renewable energy and nuclear energy as well. In terms of solar energy, for example, the improvement of photoelectric conversion efficiency and photothermal conversion efficiency will be the focus. Of course, the concrete technological solutions may be varied, for example, by changing the morphology of the surface of solar cells and through the reflection, the entire spectrum can be absorbed more efficiently; by developing multi-layer functional thin-films for transmission and absorption; or by introducing of nanotechnology and quantum control technology, *etc*. Quantum control research used to limit mainly to the solution to information functional materials. This is surely too narrow. In the future, this research is expected to be extended to the energy issue or energy-based basic research in cutting-edge areas.

In terms of computing science, we must be confident to forecast its future development instead of simply following suit as we used to. This is a possibility rather than wild fancies. Information scientists, physicists and biologists could be engaged in the forward-looking research. In 2007, the Nobel Physics Prize was awarded to the discovery of colossal magneto-resistance, which was, however, made some 20 years ago. Today, this technology has already been applied to hard disk store. Our conclusion made, at this stage, is that: it is possible to make long-term and unconventional S&T predictions, and so is it to work out China's S&T roadmap in view of long-term strategies, for example, by 2020 as the first step, by 2030 or 2035 as the second step, and by 2050 as the maximum.

This possibility may also apply to other areas of research. The point is to emancipate the mind and respect objective laws rather than indulging in wild fancies. We attribute our success today to the guidelines of emancipating the mind and seeking the truth from the facts set by the Third Plenary Session of the 11th Central Committee of the Communist Party of China in 1979. We must break the conventional barriers and find a way of development fitting into China's reality. The history of science tells us that discoveries and breakthroughs could only be made when you open up your mind, break the conventional barriers, and make foresighted plans. Top-down guidance on research with increased financial support and involvement of a wider range of talented scientists is not in conflict with demand-driven research and free discovery of science as well.

## Necessity of CAS Research on China's S&T Roadmap to 2050

Why does CAS launch this research? As is known, CAS is the nation's highest academic institution in natural sciences. It targets at making basic, forward-looking and strategic research and playing a leading role in China's science. As such, how can it achieve this if without a foresighted view on science and technology? From the perspective of CAS, it is obligatory to think, with a global view, about what to do after the 3rd Phase of the Knowledge Innovation Program (KIP). Shall we follow the way as it used to? Or shall we, with a view of national interests, present our in-depth insights into different research disciplines, and make efforts to reform the organizational structure and system, so that the innovation capability of CAS and the nation's science and technology mission will be raised to a new height? Clearly, the latter is more positive. World science and technology develops at a lightening speed. As global economy grows, we are aware that we will be lagging far behind if without making progress, and will lose the opportunity if without making foresighted plans. S&T innovation requires us to make joint efforts, break the conventional barriers and emancipate the mind. This is also what we need for further development.

The roadmap must be targeted at the national level so that the strategic research reports will form an important part of the national long-term program. CAS may not be able to fulfill all the objectives in the reports. However, it can select what is able to do and make foresighted plans, which will eventually help shape the post-2010 research priorities of CAS and the guidelines for its future reform.

Once the long-term roadmap and its objectives are identified, system mechanism, human resources, funding and allocation should be ensured for full implementation. We will make further studies to figure out: What will happen to world innovation system within the next 30 to 50 years? Will universities, research institutions and enterprises still be included in the system? Will research institutes become grid structure? When the cutting-edge research combines basic science and high-tech and the transformative research integrates the cutting-edge research with industrialization, will that be the research trend in some disciplines? What will be the changes for personnel structure, motivation mechanism and upgrading mechanism within the innovation system? Will there be any changes for the input and structure of innovation resources? If we could have a clear mind of all the questions, make foresighted plans and then dare to try out in relevant CAS institutes, we will be able to pave a way for a more competitive and smooth development.

Social changes are without limit, so are the development of science and technology, and innovation system and management as well. CAS must keep moving ahead to make foresighted plans not only for science and technology, but also for its organizational structure, human resources, management modes, and resource structures. By doing so, CAS will keep standing at the forefront of science and playing a leading role in the national innovation system, and even, frankly speaking, taking the lead in some research disciplines in the world. This is, in fact, our purpose of conducting the strategic research on China's S&T roadmap.

Prof. Dr.-Ing. Yongxiang Lu
President of the Chinese Academy of Sciences

# Preface to the Roadmaps 2050

CAS is the nation's think tank for science. Its major responsibility is to provide S&T consultations for the nation's decision-makings and to take the lead in the nation's S&T development.

In July, 2007, President Yongxiang Lu made the following remarks: "In order to carry out the Scientific Outlook of Development through innovation, further strategic research should be done to lay out a S&T roadmap for the next 20–30 years and key S&T innovation disciplines. And relevant workshops should be organized with the participation of scientists both within CAS and outside to further discuss the research priorities and objectives. We should no longer confine ourselves to the free discovery of science, the quantity and quality of scientific papers, nor should we satisfy ourselves simply with the Principal Investigators system of research. Research should be conducted to address the needs of both the nation and society, in particular, the continued growth of economy and national competitiveness, the development of social harmony, and the sustainability between man and nature. "

According to the Executive Management Committee of CAS in July, 2007, CAS strategic research on S&T roadmap for future development should be conducted to orchestrate the needs of both the nation and society, and target at the three objectives: the growth of economy and national competitiveness, the development of social harmony, and the sustainability between man and nature.

In August, 2007, President Yongxiang Lu further put it: "Strategic research requires a forward-looking view over the world, China, and science & technology in 2050. Firstly, in terms of the world in 2050, we should be able to study the perspectives of economy, society, national security, eco-environment, and science & technology, specifically in such scientific disciplines as energy, resources, population, health, information, security, eco-environment, space and oceans. And we should be aware of where the opportunities and challenges lie. Secondly, in terms of China's economy and society in 2050, we should take into consideration of factors like: objectives, methods, and scientific supports needed for economic structure, social development, energy structure, population and health, eco-environment, national security and innovation capability. Thirdly, in terms of the guidance of Scientific Outlook of Development on science and technology, it emphasizes the people's interests and development, science and technology, science and economy, science and society, science and eco-environment, science and culture, innovation and collaborative development. Fourthly, in terms of the supporting role of research in scientific development, this includes how to optimize the economic structure and boost economy, agricultural development, energy structure, resource conservation, recycling economy, knowledge-based society, harmonious coexistence between man and nature, balance of regional development, social harmony, national security, and international cooperation. Based on these, the role of CAS will be further identified."

Subsequently, CAS launched its strategic research on the roadmap for priority areas to 2050, which comes into eighteen categories including: energy, water resources, mineral resources, marine resources, oil and gas, population and health, agriculture, eco-environment, biomass resources, regional development, space, information, advanced manufacturing, advanced materials, nano-science, big science facilities, cross-disciplinary and frontier research, and national and public security. Over 300 CAS experts in science, technology, management and documentation & information, including about 60 CAS members, from over 80 CAS institutes joined this research.

Over one year's hard work, substantial progress has been made in each research group of the scientific disciplines. The strategic demands on priority areas in China's modernization drive to 2050 have been strengthened out; some core science problems and key technology problems been set forth; a relevant S&T roadmap been worked out based on China's reality; and eventually the strategic reports on China's S&T roadmap for eighteen priority areas to 2050 been formed. Under the circumstance, both the Editorial Committee and Writing Group, chaired by President Yongxiang Lu, have finalized the general report. The research reports are to be published in the form of CAS strategic research serial reports, entitled *Science and Technology Roadmap to China 2050: Strategic Reports of the Chinese Academy of Sciences*.

The unique feature of this strategic research is its use of S&T roadmap approach. S&T roadmap differs from the commonly used planning and technology foresight in that it includes science and technology needed for the future, the roadmap to reach the objectives, description of environmental changes, research needs, technology trends, and innovation and technology development. Scientific planning in the form of roadmap will have a clearer scientific objective, form closer links with the market, projects selected be more interactive and systematic, the solutions to the objective be defined, and the plan be more feasible. In addition, by drawing from both the foreign experience on roadmap research and domestic experience on strategic planning, we have formed our own ways of making S&T roadmap in priority areas as follows:

### (1) Establishment of organization mechanism for strategic research on S&T roadmap for priority areas

The Editorial Committee is set up with the head of President Yongxiang Lu and the involvement of Chunli Bai, Erwei Shi, Xin Fang, Zhigang Li, Xiaoye Cao and Jiaofeng Pan. And the Writing Group was organized to take responsibility of the research and writing of the general report. CAS Bureau of Planning and Strategy, as the executive unit, coordinates the research, selects the scholars, identifies concrete steps and task requirements, sets forth research approaches, and organizes workshops and independent peer reviews of the research, in order to ensure the smooth progress of the strategic research on the S&T roadmap for priority areas.

### (2) Setting up principles for the S&T roadmap for priority areas

The framework of roadmap research should be targeted at the national level, and divided into three steps as immediate-term (by 2020), mid-term (by 2030) and long-term (by 2050). It should cover the description of job requirements, objectives, specific tasks, research approaches, and highlight core science problems and key technology problems, which must be, in general, directional, strategic and feasible.

### (3) Selection of expertise for strategic research on the S&T roadmap

Scholars in science policy, management, information and documentation, and chief scientists of the middle-aged and the young should be selected to form a special research group. The head of the group should be an outstanding scientist with a strategic vision, strong sense of responsibility and coordinative capability. In order to steer the research direction, chief scientists should be selected as the core members of the group to ensure that the strategic research in priority areas be based on the cutting-edge and frontier research. Information and documentation scholars should be engaged in each research group to guarantee the efficiency and systematization of the research through data collection and analysis. Science policy scholars should focus on the strategic demands and their feasibility.

### (4) Organization of regular workshops at different levels

Workshops should be held as a leverage to identify concrete research steps and ensure its smooth progress. Five workshops have been organized consecutively in the following forms:

**High-level workshop on S&T strategies.** Three workshops on S&T strategies have been organized in October, 2007, December, 2007, and June, 2008, respectively, with the participation of research group heads in eighteen priority areas, chief scholars, and relevant top CAS management members. Information has been exchanged, and consensus been reached to ensure research directions. During the workshops, President Yongxiang Lu pinpointed the significance, necessity and possibility of the roadmap research, and commented on the work of each research groups, thus pushing the research forward.

**Special workshops.** The Editorial Committee invited science policy scholars to the special workshops to discuss the eight basic and strategic systems for China's socio-economic development. Perspectives on China's science-driven modernization to 2050 and characteristics and objectives of the eight systems have been outlined, and twenty-two strategic S&T problems affecting the modernization have been figured out.

**Research group workshops.** Each research group was further divided into different research teams based on different disciplines. Group discussions, team discussions and cross-team discussions were organized for further research, occasionally with the involvement of related scholars in special topic discussions. Research group workshops have been held some 70 times.

**Cross-group workshops.** Cross-group and cross-disciplinary workshops were organized, with the initiation by relative research groups and coordination by Bureau of Planning and Strategies, to coordinate the research in relative disciplines.

**Professional workshops.** These workshops were held to have the suggestions and advices of both domestic and international professionals over the development and strategies in related disciplines.

### (5) Establishment of a peer review mechanism for the roadmap research

To ensure the quality of research reports and enhance coordination among different disciplines, a workshop on the peer review of strategic research on the S&T roadmap was organized by CAS Bureau of Planning and Strategy, in November, 2008, bringing together of about 30 peer review experts and 50 research group scholars. The review was made in four different categories, namely, resources and environment, strategic high-technology, bio-science & technology, and basic research. Experts listened to the reports of different research groups, commented on the general structure, what's new and existing problems, and presented their suggestions and advices. The outcomes were put in the written forms and returned to the research groups for further revisions.

### (6) Establishment of a sustained mechanism for the roadmap research

To cope with the rapid change of world science and technology and national demands, a roadmap is, by nature, in need of sustained study, and should be revised once in every 3–5 years. Therefore, a panel of science policy scholars should be formed to keep a constant watch on the priority areas and key S&T problems for the nation's long-term benefits and make further study in this regard. And hopefully, more science policy scholars will be trained out of the research process.

The serial reports by CAS have their contents firmly based on China's reality while keeping the future in view. The work is a crystallization of the scholars' wisdom, written in a careful and scrupulous manner. Herewith, our sincere gratitude goes to all the scholars engaged in the research, consultation and review. It is their joint efforts and hard work that help to enable the serial reports to be published for the public within only one year.

To precisely predict the future is extremely challenging. This strategic research covered a wide range of areas and time, and adopted new research approaches. As such, the serial reports may have its deficiency due to the limit in knowledge and assessment. We, therefore, welcome timely advice and enlightening remarks from a much wider circle of scholars around the world.

The publication of the serial reports is a new start instead of the end of the strategic research. With this, we will further our research in this regard, duly release the research results, and have the roadmap revised every five years, in an effort to provide consultations to the state decision-makers in science, and give suggestions to science policy departments, research institutions, enterprises, and universities for their S&T policy-making. Raising the public awareness of science and technology is of great significance for China's modernization.

Writing Group of the General Report

February, 2009

# Preface

Manufacturing industry covers a variety of industries and is the main industry of the national economy. China has become a major manufacturing country, but not a manufacturing great power. So it is even imperative to carry out advanced manufacturing research in China. Therefore, Chinese Academy of Sciences organized its experts and carried out research on the developing roadmap of advanced manufacturing technology.

Compared with the traditional manufacturing technology, advanced manufacturing technology has its remarkable features and more extensive meanings. The development direction of advanced manufacturing technology has been discussed continuously. In general, it is believed that globalization, informationization, intelligentization, greening and multi-disciplinary integration is the developing direction of advanced manufacturing technology. This report focuses only on two main development directions of advanced manufacturing, i.e., green manufacturing based on ubiquitous information and intelligent manufacturing based on environmental friendliness.

Tianran Wang, Haibin Yu, Feiyue Wang, Yunlong Zhu, Wensheng Zhang, Hong Song, Peng Zeng, Tao Ku, Fen Xia, Xiong Gang, Yanwu Yang, and Dong Yu participated in the writing of intelligent manufacturing based on ubiquitous sensing (perception). Yi Zhang, Huiquan Li, Shili Zheng, Hongzhang Chen, Xiangping Zhang, Chao Yang, Hongbin Cao, Xiaowei Peng participated in the writing of environmentally friendly green manufacturing. Tingcan Ma offered a great deal of information for the topic research. Tianran Wang, Yunlong Zhu and Hong Song made all draft together.

During the report compiling and writing process we received strong support from Jiaofeng Pan, Director General, and Feng Zhang, Section Chief of Bureau of Strategic Planning of the Chinese Academy of Sciences (CAS), and also got help from Xuemin Wu, researcher with Shenyang Institute of Automation. Hereon we would like to express our gratitude to them.

Since strategic research is a dynamic process, this report is inevitable to be imperfect and inaccurate due to the limitation of knowledge and time of the research group. Comments and corrections are welcomed.

Research Group on Advanced Manufacturing
Technology of the Chinese Academy of Sciences
May, 2010

# Contents

# Abstract

Manufacturing industry is the material basis of the national economy and the main industry which occupies an important position in the national economy. China is a big manufacturing country, but not a manufacturing great power, therefore developing advanced manufacturing technology has particularly great significance for the development of China's national economy.

Manufacturing industry includes 31 departments, whose output is the main component of industrial gross output. Compared with traditional manufacturing technology, advanced manufacturing technology concerns not only manufacturing process but also market analysis, product design, machining, assembly, sales, maintenance, services and recycle.

Advanced manufacturing technology is an important and broad study field. It has been paid more attention in China and widely applied in the research and development of major equipments. Large amount of research work has been carried out in recent years.

People have been continuously discussing on the development direction of advanced manufacturing technology. Basically, they believe that the development direction of advanced manufacturing technology will be the globalization, informationization, intelligentization, greening and integration of multi-disciplines. "Looking at the evolution trends of advanced manufacturing technology, 'green' and 'intelligence' will be its main direction of development."

In the report, intelligent manufacturing based on ubiquitous information and green manufacturing in harmony with environment are studied.

With the development and maturity of such technologies as industrial wireless networks, sensor networks, radio frequency identification and micro-electro–mechanical systems, the comprehending process of people on manufacturing technology develops from the "insufficient understanding" of manufacturing equipment and manufacturing processes, to the multi-dimensional (three-dimensional space plus time) and transparent ubiquitous sensing, which contributes to the coming of "ubiquitous information manufacturing" era–the technology of new generation information manufacturing and automation based on ubiquitous sensing with ubiquitous information technology as the main driving force. This development will also become the core driving force for the advancement of new generation advanced manufacturing technology. It is foreseeable that the new generation ubiquitous information manufacturing systems will greatly improve the manufacturing efficiency and product quality of traditional manufacturing mode, reconstruct enterprise organizations and business processes, innovate enterprise operation mode, greatly reduce product cost and resource consumption, consequently provide user with more transparent and personalized service, and will eventually become a developing direction of intelligent manufacturing based on man-machine harmony and ubiquitous information .

In the next five years, with the application of industrial wireless network, sensor network, new sensors, RFID, etc., the embryonic forms and typical applications of ubiquitous information manufacturing will appear. In the next decade, the industrial control system with ubiquitous sensing ability will be constructed, and the ubiquitous sensing will include information perception and its utilization methods, massive information processing, control management methods and the corresponding standards. This will have an inestimable impact on various fields of manufacturing industry, and will eventually achieve the transition of manufacturing system from human-machine harmony to intelligent manufacturing era of autonomous operation with machine as the main part.

Green manufacturing can efficiently utilize resources and energy without pollution and cover full life cycle of products concerning design, machining, utilization and recycle by innovating traditional manufacturing technology, design concept and production methods.

Green process is a big technology system. And the system takes resource and environment as the guidance, utilizes the concept of atom economy in substance transformation, applies the principles of species symbiosis, substance regenerative cycle and ecological integrity in natural ecology, combines with the highly effective hierarchical and multistage substance utilization of systems engineering and optimization methods design, and gives full play to the potential of material resources to reduce waste from the source. It includes the atom economic reaction of highly-effective and clean resource transformation and the green design and process intensification of separation process on the microscopic scale; the process coupling and adjustment, the optimized integration of material, energy, and information flows, and environment-oriented multi-objective optimization on the synoptic scale; and the system integration on large-scale system of ecological industry group. Green process can realize the global optimization and support the sustainable development of manufacturing industry.

As to the greening technology upgrading of the process manufacturing industry for large-scale resource transformation and utilization of minerals, oil and biomass, innovative breakthrough should be made in the green process and engineering of substance transformation and in the engineering direction of new energy, new resource alternative technology and biomass processing and refining of process industry. By 2030, in the key directions such as the new reaction medium alternative technologies, highly-efficient catalytic technology, process intensification & advanced reaction separation equipment, core technologies of resource recycling and environment, etc., green process engineering and technology will be established to substantially increase resource utilization efficiency, reduce energy consumption and decrease waste discharge. As to discrete manufacturing industry, the product green design and life cycle evaluation systems will be used widely, and non-waste machining technology and recycling and reconstruction technology for mechanical and electrical products and automotive products will be popularized. By 2030, the manufacturing process will reduce 50% of its energy consumption and raw material loss rate; its secondary resource recycling and utilization rate will reach 70% and the chemical environmental risks will be basically controlled. By 2050, the raw material loss rate of the process manufacturing can be further reduced by 90%, atom economy and selectivity of chemical conversion will be close to 100%, zero discharge of waste will be realized, and chemical environmental risks will be completely eliminated. The production system with arterial and venous integration and the advanced manufacturing system with new low-carbon economy model will be formed, greening of the process manufacturing industry will be realized, and China will enter the ranks of the advanced countries in the world.

# Introduction

## 1. Manufacturing Industry Is an Important Area of National Economy

Manufacturing industry covers a broad area. According to the regulation of the *National Economic Industry Classification* (GB/T 4754-2002), the national standard of China, manufacturing industry contains 31 categories, such as iron and steel, nonferrous metals, clothing, petroleum chemical industry, equipment manufacturing, agro-food processing industry, textile, as well as waste resources and waste materials, recycling and processing, etc.

Based on the manufacturing process, manufacturing industry can be divided into two categories of continuous manufacturing and discrete manufacturing. A typical continuous manufacturing includes metallurgy, chemical industry, papermaking and so on, while machinery manufacturing is a typical discrete manufacturing. In addition, some manufacturing processes belong to semi-continuous and semi-discrete manufacturing.

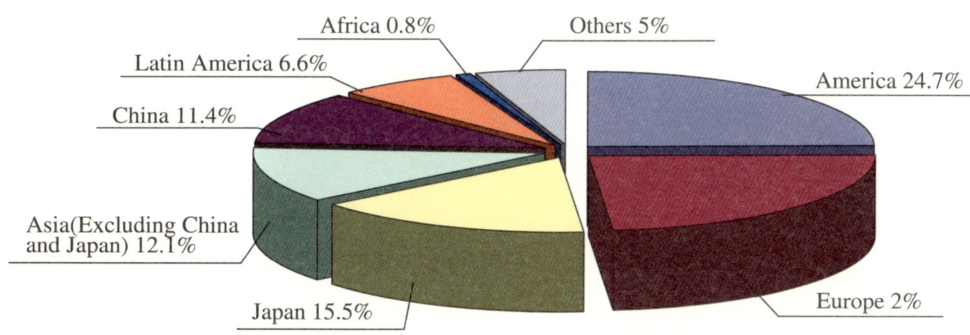


Figure 0.1　World Manufacturing Value Added (MVA) Share in 2007 (Note: Estimated according to the 2000 constant price. Source: http://cpc.people.com.cn/GB/64093/82429/83083/7164596.html)

Manufacturing industry is the material basis of the national economy and main industrial body, the engine for the rapid growth of economy and an important guarantee for national security, it is also the basic carrier of science and technology, and an important manifestation of the national economy and overall national power. Manufacturing industry occupies an important position in the national economy for both developing and developed countries; it is always an important and strategic component of the national economy. According to statistical data of 2005, the manufacturing added value of the

United States, Japan and Germany accounted for14%, 21% and 23% of their GDPs respectively, moreover this proportion is still growing in 2006. Industrial added value of China's manufacturing industry is 7.261956 trillion Yuan, and occupies 34.4% of the GDP, 79.53% of all industry. It is obvious that China's manufacturing industry has a great impact on the national economy. Figure 1 shows the ranking of global manufacturing in 2007.

China was once a big manufacturing country in history. In the 15th century, China's science and technology was among the first class in the world, China was a big manufacturing country until the 17th century. After the 17 centuries, the situation became worse and China became the prey of the power countries. Since the reform and opening up, China's manufacturing industry has made rapid development. From 1990 to 2005, the annual growth rate of manufacturing added value of the United States, Japan, Germany and China were 2.07%, 0.65%, 1.38% and 7.79% respectively, it shows that the growth rate of China's manufacturing added value is far higher than those of the other three countries. From the view of the manufacturing scale, China's manufacturing added value was 759.661 billion U.S. dollars in 2005, ranking the third in the world, which surpassed the manufacturing added value of Germany and was equal to 79.79 % and 43.70% of the manufacturing added values of Japan and the United States respectively, and China became a big manufacturing country.

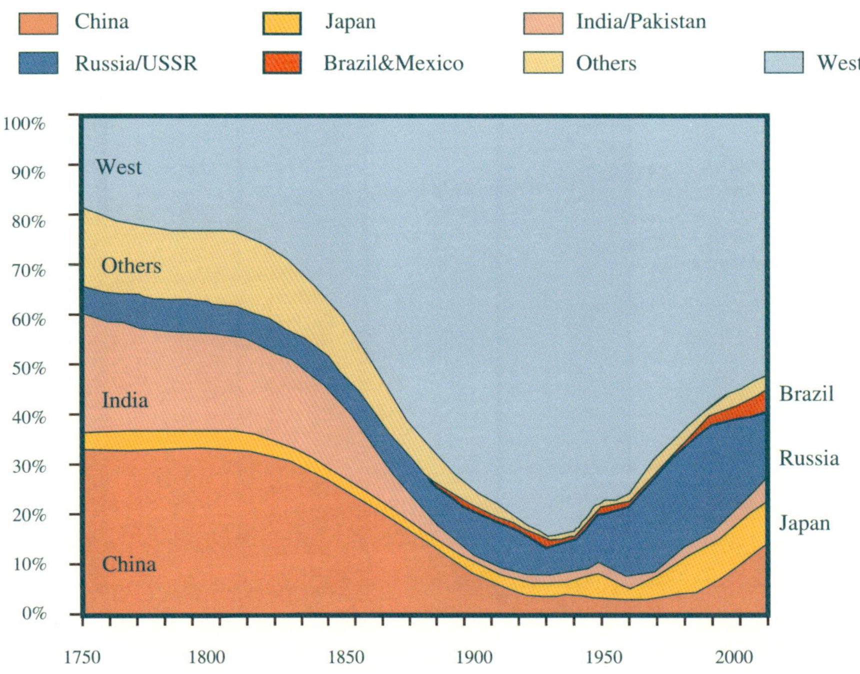


Figure 0.2　China was once a big manufacturing country in the history. (Source：Assuring the future of manufacturing in Europe, report of the High-Level Group September 2006)

As we all know that China is not a powerful manufacturing country yet. According to the data of 2004, China's manufacturing added value rate was 27.5%, while the United States and Japan were 41.2% and 36.9% respectively. This shows that China's manufacturing industry is still in the low level of profitability relatively, and the economic efficiency needs improving urgently. The product of manufacturing industry in China is in the low end; manufacturing technology for major equipment, specially the key technologies relies heavily on foreign countries; China is still in the low technical and low added value level in the international industrial division; the utilization rate of energy and material is low, product energy consumption in the major industry is 40% higher than that of the international advanced level. Therefore, advanced manufacturing technology needs developing urgently in China.

## 2. Development of Advanced Manufacturing Technology

Manufacturing technology can be defined as a series of technologies used to transform raw materials into products.

Combining tradition manufacturing technology with information and other technologies forms the advanced manufacturing technology. Breakthroughs in the fields of information, material technologies, etc. greatly promote the development of tradition manufacturing technology, which attract the concerns from scientist. In 1980s, Americans has firstly coined the noun of advanced manufacturing technology (AMT) in their National Research Program.

Compared with the traditional manufacturing technology, advanced manufacturing technology has its remarkable features and wide meanings. Advanced manufacturing technology concerns not only manufacturing process but also market analysis, product design, machining, assembly, sales, maintenance, services and recycle.

Advanced manufacturing technology is divided into three parts:
- Modern design technology
- Advanced manufacturing technology and equipment
- Systems management technology

The purpose to develop advanced manufacturing technology is not only to produce high quality products efficiently to meet the needs of users, but also to improve product adapting ability and competitiveness in the dynamically changing market with the goal of low consumption, clean and flexible production.

The discussion concerning the development direction of advanced manufacturing technology never stops. It is believed that globalization, informationization, intelligentization, greening and multi-disciplinary integration will be the developing direction of advanced manufacturing technology.

"Looking through the evolution trends of advanced manufacturing technology, 'green' and 'intelligence' will be its main direction of development."

Advanced manufacturing technology is listed as an important research field and given strong support by the Chinese government. This makes advanced manufacturing technology develop rapidly and promotes the  upgrade of manufacturing capacity and the reform of production organization and management in China. This will also impel China to stride forward to a big and powerful manufacturing country.

In order to become a powerful manufacturing nation, in the next 10-15 years, China should improve its independent design and manufacturing capacity of major and important equipment, and enhance the utilizing rate of material and energy. To make breakthroughs in the design and manufacturing technology of major equipment, to economize the energy and raw material, and to protect the environment will be the key point of development of advanced manufacturing technology in this stage.

In addition, many aspects concerning modern design technology, advanced manufacturing technology and equipment, and systems management technology will be developed. For example, precision and ultra-precision machining technology, laser processing technology, near net shape technology and so on in the machinery manufacturing field of discrete manufacturing industry; research on the elimination of "capital-intensive", "energy intensive" and "material consumption-intensive" features in the chemical industry of the continuous process industry; chemical molecular design and advanced synthesis technology, process intensification, industrial biotechnology and so on in the chemical products and materials, All these technologies will be important development directions in a long time.

Advanced manufacturing technology is the technology integration of the related technology fields such as traditional manufacturing technology and information technology, etc. which in the subsequent developments will be continuously combine with information, materials, biology, and nanotechnology and MEMS technology with faster speed to impel its development. The development of information technology, materials technology and biotechnology will also supplement new content for advanced manufacturing technology continuously.

Information technology will promote the development of manufacturing technology continuously in the control and transformation of the manufacturing equipment and in the management of production systems, and will push manufacturing technology towards to the development of intelligent manufacturing.

With the development of sensing technology and wireless sensor network technology, the manufacturing environment will become more transparent, manufacturing will enter the sensing era of the rich-information from limited information, and production management and control of the production process will be changed

Continuous emergence of advanced materials will greatly enhance the efficiency and cleanliness of process industry, will improve the performance of machining equipment, and will produce new equipment.

   *Advanced Manufacturing Technology in China: A Roadmap to 2050*

Advances in brain science will bring revolution in the way of human-computer interaction. People will cooperate with the increasingly intelligent machines to develop people's skills and creativity. Co-ordination between man and machine will be the characteristics and challenges of future manufacturing systems.

With the progress of manufacturing technology, in order to meet the needs of different users in the competition, a new concept of production management will be proposed, such as service-oriented manufacturing, and reliable manufacturing, etc. It is predicted that in the future it will play a leading role that production system of man-machine cooperation to exert their respective functions, and Taylor's theory that once played an important role in the history will be updated.

Scientists from home and abroad have carried out comprehensive study in the resources green transformation, new resources and energy substitution, process intensification and system integration, green product engineering in the process industry, as well as raw materials selection, design, manufacturing, packaging and recycling area in the discrete manufacturing. Green manufacturing has become an important feature of the manufacturing industry in the 21st century.

In the next 30 years, China's main manufacturing equipment and technology will be developed with the world step, micro-nano-manufacturing, bio-manufacturing, and other newly emerging technologies will become the major industry and support other related industries. Man-machine harmony manufacturing theory and technology will be set forth and applied. Green manufacturing, a symbol of technological progress, will evolve from the conception and objective to system technology and production standards , and become a symbol of technological progress. With the development of information technology, new material technology, biology and brain science and technology, and nano-technology, the development of manufacturing technology itself will be greatly affected. New manufacturing technology contributing to the technological advances in these areas will become the symbol of advancement in advanced manufacturing technology in the future.

By 2050, comprehensive promotion of intelligentization and greening on manufacturing industry will be completed in China, and high efficient and clean recycling use of resources and energy and the minimum environmental effect will be achieved. With international first–class equipment, innovative design and manufacturing capabilities, the intelligent and green manufacturing industry system will be constructed in China.

| | | Around2020 | Around2030 | Around2050 |
|---|---|---|---|---|
| Advanced manufacturing | Level of manufacturing industry | Core technology dependent on foreign country is less than 30% | Core technology dependent on foreign country is less than 20% | Core technology dependent on foreign country is less than 50% |
| | Manufacturing equipments | Basically change the situation of major equipments dependent on import | Meet the needs of development and manufacturing for major equipments | Possess the world-class capability of equipment innovation and manufacturing |
| | Manufacturing intelligence | Widely apply automatic manufacturing based on ubiquitous awareness so that increase more than 10% of productivity | Establish intelligent system and management manufacturing system with harmonious man-machine | Possess manufacturing system using intellitent machines and self-control |
| | Products green design | Dismountable and recycle of mechanical and electronis products and automotive products | Establish full cycle green design criterion for major products | Popularize full cycle green design and recycle of products |
| | Energy Conservation and reductions in carbon emissions | Manufacturing process decreasing 30% and carbon emissions reducing 20% | Manufacturing process decreasing 50% and carbon emissions reducing 30% | Found manufacturing system with low carbon economy |
| | The efficient clean utilization of resources | Raw material loss rate will be reduced by 30%,recycling utilization rate of secondary resources will reach up to 50% | Raw material loss rate will be reduced by 50%,recycling utilization rate of secondary resources will reach 70% | Raw material loss rate will be reduced by 90%,recycling utilization rate of wastes will reach 90% |
| | Environmental impact | Environmental pollution of manufacturing process will be basically under control | Nearly zero dischage of hazardous waste,chemical environment risk will be under control basically | Zero discharge of hazardous waste,chemical environmental risks will eliminated |

## 3. Intelligent Manufacturing based on Ubiquitous Information

People dominate and carry out manufacturing so as to create material wealth for themselves. In the manufacture process, people developed from using simple tool to using complex tool, and further developed to using machines. During the late 18th century, the Industrial Revolution characterized by the invention of steam engine and machine tools, a new era of manufacturing industry dominated by machines began. Machines replaced physical labor of human being, and productivity and production efficiency were greatly improved. How to use and control the machine, and how to manage the production system consisting of machines became the core concern in the manufacturing filed. In the late 19th century, production line and mass production mode were founded by Ford. Subsequently Taylor created the scientific management theory, where functions of manufacturing system were categorized, manufacturing technology was divided specifically, therefore the new era of production organization and management was opened up.

Throughout the manufacturing process, information has always been an indispensable and important element. In the early time, people estimated the machining process with their observations and sensation. In the mass production era, with the increasing of product processing accuracy and speed, processing tools developed into machines and production systems, and testing equipment became the origin source of control processing information. Timely and accurate information acquisition can support the development of automatic control system, while with the development of automatic control systems, production efficiency and product quality are increased continuously. With the integration and development of the information technology and industrial applications technology, a number of new manufacturing modes were proposed and applied, such as lean production, agile manufacturing, virtual manufacturing, virtual enterprises, etc. And these patterns allow the transformation from technology-centered mode to people-centered mode, from the pyramid-like multi-level production and management structure to a flat network structure, from the traditional serial work mode to the parallel working mode, from the fixed organization form which divided departments according to the functions to group forms of work organization of dynamic, self-management; from compliance quality to the satisfaction of quality.

Information is very important, however, information we can get now is not sufficient. In the running process of machines, due to the lack of inspecting and testing means for machine's own behavior and environmental change (abrasion wear, temperature changes, etc.), the error can not be adjusted automatically. In adequate information support leads to large amount of costs in the equipment commissioning, program preparation and debugging test and causes low utilization rate of machines. Insufficiency of relevant information also influences the organization of production and management efficiency as well as automation.

With the maturity of industrial wireless networks, wireless sensor network (WSN), radio frequency identification (RFID), micro-electrical mechanical systems (MEMS), etc., the technical level of information acquisition, identification, processing, transmission, searches, analysis and utilization have been greatly improved in various fields, the capacity of people to obtain, control and use information will be greatly expand, and the manufacturing industry will face the situation in which manufacturing information becomes more abundant gradually, making people from "lack of understanding" of manufacturing equipment and manufacturing processes at present, to ubiquitous sensing development of multi-dimensional, and transparence of three-dimensional space plus time. Informationized manufacturing based on ubiquitous sensing will be our new opportunities and challenges, which will contribute to the coming of the era of information manufacturing and automation technology based on ubiquitous sensing — "ubiquitous information manufacturing". The future manufacturing environment is a physical space which embedds sensing, computing and control information equipment; through the collaborative action of the network distributed intelligent devices, users can easily obtain information of physical space, and implement the corresponding decision by means of intelligent equipment. Virtual reality, especially real three-dimensional display technology will change traditional design, manufacturing and assembly mode. Space collaboration technology will enhance the agility, adaptability and production efficiency of manufacturing system in the background of integrated applications of manufacturing system. Parallel management technology will integrate management, people, equipment, technology and other factors to achieve parallel implementation of the actual manufacturing system. E-commerce technology will deeply integrate manufacturing and marketing processes of advanced manufacturing, and enhance the core competitiveness ability of enterprises, as well as the ability of adapting to the complex dynamic market. It is foreseeable that new generation of ubiquitous information manufacturing systems will greatly increase manufacturing efficiency and product quality, reconstruct enterprise organizations and business processes, innovate operation mode of enterprises, greatly reduce product cost and resource consumption, and provide more transparent technology and personalized service for users. "Ubiquitous information manufacturing" will become an important developing direction of advanced manufacturing.

Some countries have started researches on single technology in the ubiquitous information manufacturing, and proposed clear and positive plan for national industrial development.

In the next five years, with the applications of the industrial wireless network, sensor network, new sensors, RIFID and so on, the rudiment of ubiquitous information manufacturing and typical applications will be formed; In the next decade, ubiquitous sensing, intelligent manufacturing collaboration space, sensor networks and industrial wireless networks, human-computer interaction, parallel management, and three-dimensional display, system technology and e-commerce will support the development of manufacturing towards the direction of intelligent manufacturing; and the industrial control system of ubiquitous sensing including the perception and use means, mass information processing, control management methods and the corresponding

   *Advanced Manufacturing Technology in China: A Roadmap to 2050*

standards will be constituted.

Intelligent manufacturing not only applies human experience and skill to the automatic control of equipment and production systems, but also include the sensation and adjusting components and the use of intelligent materials.In particular, the improvement of perceived capabilities, will make self-adjusting of manufacture system for its own behavior become possible. Forecasting for the next 20 years, the level of equipment with amenity will be greatly improved and intelligent equipment with the self-adaptive, self-optimizing, self-calibration, self-diagnosis and self-repair will continuously emerge. The appearance of these new technologies and innovation of information technology will enable manufacturing systems of new generation to show the situation of man-machine cooperative work with the human decision-making as main body, more personalized, intelligent and friendly in the next 20 years. With ubiquitous network as the basis, ubiquitous sensing as the core, ubiquitous service as the goal, the ubiquitous intelligence as expansion and enhancement, the integrated production organization and scheduling, equipment management and cooperation, and further reasonable and precise control of industrial production activities, will have an inestimable impact on various fields of the manufacturing industry.

## 4. Green Manufacturing Harmony with the Environment

The use of natural resources and energy and the use of tools and machinery promote the development and innovation of the manufacturing industry.

Manufacturing is to create wealth for human with the use of natural resources. The ancient human made stone tools, pottery, smelt metal, and gunpowder, and these activities all belong to the development of natural resources. In manual production, the utilization of natural resources is limited in manufacturing process, and its influences on environmental change are slow.

Modern large-scale mechanized productions provide unprecedented material wealth for human being. Since the industrial revolution, not only the productive capacity of human being increased rapidly and resources were consumed greatly, but also the chemical industry entered quickly into industrial and agricultural economy. Development of petrochemical industry from small-scale inorganic chemical industry in the mid-18th century to the production of isopropyl alcohol with propylene in 1920 has greatly improved and enriched the material life of human being. However, the resources and environmental capacity of natural ecological system are limited, which leads and contradictions between stability mechanism of natural ecological system and the limitless growth mechanisms of human social economy, will inevitubly become increasingly intensified. As Engels warned us long ago: "We should not be over enjoyed oure victory on nature. For each such victory, the nature will take revenge on us. Each victory, initially did achieve our desired results, but the next, and then further on, something completely different and unexpected effects occurs, which often eliminated the initial benefits." After paying out a painful price, human beings began to think, and then choose the economic mode of sustainable development.

Green manufacturing can efficiently utilize resources and energy without pollution and cover full life cycle of products concerning design, machining, utilization and recycle by innovating traditional manufacturing technology, design concept and production methods.

The green process takes the resources and environment as the guidance, utilizes the atomic efficient concept of substance transformation and the natural ecology species coexistence, the material regenerative cycle and the ecology conformity principle, combining systems engineering and the substance highly effective lamination multistage uses of the optimization methods design, and fully utilizes resources potential to realize the big technology system of reducing the waste from the origin source. It containes the atomic efficiency reaction of resources highly effective clean transformation on the microscopic scale and the green design and the process intensification of the separation process, the process coupling and adjustment controlling, the material flow – energy flow – information flow's optimization integration and the multi-objectives optimization guided by the environment, the system integration of ecology industry group large-scale system level, in order to realized the overall optimization and provided the support for the manufacturing industry sustainable development.

Green manufacturing technology research will be carried out in the following four aspects: the green transformation of resources, new resources and energy substitution, process intensification and systems integration as well as green products manufacturing. Carbon-intensive production processes and patterns with high consumption of energy, etc. will fundamentally be changed, while low-carbon economy technology including low-carbon products and processes will become technology field of keystone development gradually, improve greatly resource utilization rate, reduce energy consumption, green process engineering technology of reducing energy consumption and waster discharging will be developed rapidly, and products green design and full life-cycle assessment system will be get widely applied. Waste-free machining technology of discrete manufacturing and recycle technology of mechanical and electrical products, automotive products will be popularized.

The widespread use of new energy sources, such as renewable energy, nuclear energy will change the energy structure of the existing manufacturing. New materials, new resources substitution, technologies and biomass processing refining projects will achieve new breakthroughs, Substitution resources, such as, non-high-quality resources, biomass resources, the second resources, etc. will gradually become the main body of the manufacturing resources machining, at the same time, the requirements of resources recycle will also change the current structure form and use methods of the products.

China attaches great importance to green manufacturing, with orientation of the country's policies and regulations and the support of science and technology innovation, it is expected by 2020, energy consumption of annual manufacturing process reducing 30%, carbon emissions reducing 20%, raw material loss rate reducing 30%, secondary resources recycling utilization rate 50%, the environmental pollution of the manufacturing process will be under control.

By 2050, a low carbon economy manufacturing system will be established, to achieve zero discharge of hazardous waste, and basically eliminate of chemical environmental risks, popularize the green design and recycle of products entire process, and to achieve green manufacturing of harmony with the environment.

## 5. Conclusion

Advanced manufacturing technology covers comprehensive technical researches. This study only focuses on the developing trends of intelligent manufacturing based on ubiquitous information and green manufacturing harmony with environmental, as well as their influences on the development of advanced manufacturing technology.

# Part I

## Intelligent Manufacturing Systems Based on Ubiquitous Information

# 1 Overview

Automation of information processing and application has a great effect on promoting the development of national economy, greatly accelerates the process of industrialization, remarkably improves material living standard of people, and changes the production and operation management mode of enterprises from the past extensive management gradually to the precise management. The networking of information applications, on one hand significantly improved people's production and living standards, on the other hand has profound impact on people's production and lifestyle. With the application of e-commerce, enterprise resource planning, supply chain management, product lifecycle management, etc., the management and operation mode of enterprises have been profoundly changed, the advanced manufacturing models such as agile manufacturing, concurrent engineering, mass customization, network collaborative design and manufacturing, as well as advanced management models such as business process reengineering, flat organizational structure, learning-oriented enterprises have emerged.

With the emergence and mutual integration of new technologies characterized by perception and intelligence, the main features of future development of information technology will be ubiquitous information technology everywhere. With the maturity and development of some technologies such as RFID, sensor networks, industrial wireless networks, MEMS and sensor technology, information manufacturing and automation technology of new generation (U-manufacturing) with ubiquitous perception as representative will be a new driving force to promote the development of advanced manufacturing technology. It will make people develop from "lack of understanding" of manufacturing equipment and processes, to the multi-dimensional ubiquitous perception and transparency of the three-dimensional space plus time. Computing mode based on ubiquitous technology will naturally and deeply fuse many technologies, such as various terminals with environmental perception capability, mobile communication, information acquisition, context awareness, intelligent software and human-computer interaction, into every corner of the manufacturing industry. As the infrastructure of next generation information service, ubiquitous manufacturing information perception space will greatly increase the manufacturing efficiency and product quality, reduce the product cost and resource consumption, provide users with a more transparent and personalized service, and greatly improve the production efficiency.

Ubiquitous information manufacturing technology is a comprehensive and integrated information processing technology, with ubiquitous network as the basis, ubiquitous perception as the core, ubiquitous service as the purpose, and the expanding and upgrading of ubiquitous intelligence as the goal. The so-called ubiquitous networks mainly refer to the basic environment which integrates Internet, communications, computers, sensor networks, RFID, etc., and its impact mainly includes promoting intelligent device networking, strengthening IP basic construction, shortening the digital divide, completely changing and improving communication means of people. With the expansion of time and space dimensions management of productive activities, the real-time and traceability of logistics management has been markedly improved, and will have enormous impact on production management. Ubiquitous perception refers to a technical method which uses micro-systems and electronic devices as hardware foundation of ubiquitous information technology, by means of embedded systems with variety of application environments to capture, analyze and transmit information of multiple modes (sound, light, electricity, heat, physical information, chemical information, actions, and spatial information). It is an unprecedented technical means for human being to deeply understand the physical world. It will greatly expand people's capabilities of understanding and monitoring the physical world, and promote further rationalization and refinement control of industrial activities in various fields. With providing users with convenient, and deep information services as the goal, with some information processing technologies such as the models based on mechanism, statistical models, data mining as the means, ubiquitous service deeply mines and learns the mass data produced by ubiquitous network and ubiquitous sensing so as to  prevent information from submerging into the oceans of massive data. Ubiquitous intelligence, expands and enhances the ubiquitous information processing technologies, it is characterized by self-adaptation, self-optimizing, self-configuring, self-restore of ubiquitous information services, in order to provide plug and play, and real-time customization and highly abstract. fields knowledge for different areas.

Both historical experience and the laws of scientific and technological development will show that ubiquitous information technology will have a profound impact on manufacturing technology and manufacturing industry for quite a long time, especially it will have an inestimable impact on the fields of ultra-precision machining manufacturing with high precision demanding, and manufacturing process fields under extreme conditions which require high-temperature, high pressure, high humidity, strong magnetic field, strong corrosion and so on. The fundamental difference between the manufacturing mode based on ubiquitous information manufacturing technology and traditional manufacturing mode, is that the perception capacity of the physical world is improved fundamentally, while it will extend to two directions, one is the perception extension along the downward and another is information extension along upward, enterprise organization and management processes as well as operational modes in the macro level will have some profound changes.

E-manufacturing system based on ubiquitous perception is facing with a series of challenging problems.

(1) The establishment technologies of ubiquitous information perception space, including ubiquitous industrial network technology, equipment description technology, new generation industrial process monitoring and massive information collection capabilities and technology.

(2) The massive manufacture information processing technologies, including space aggregation of multi-dimensional information, integration technology of different physical quantities information, information integration technologies of multi-sampling rate, massive information analysis and knowledge mining techniques.

(3) The innovative manufacturing mode in the ubiquitous information perception space, including the quasi-real time, ultra-precise processing manufacturing mode, flat manufacturing mode of multi-level feedback and cross-layer collaboration, the task-oriented production line self-organization manufacturing mode.

(4) Manufacturing information and automation technologies in the ubiquitous information perception space, including real-time tracking high efficient logistics and near zero inventory technology, self-diagnosis, self-configuration health monitoring and predictive maintenance technology of equipment, feedback control technology based on multi-source, multi-dimensional information.

(5) Cognition ability of machine for human behavior and new human-machine interaction technology.

The integration of ubiquitous network and sensor technology will greatly enhance the human ability to interact with the physical world. Ubiquitous information perception space will have a revolutionary impact on the manufacturing industry. In the next 20 years, e-manufacturing system technology based on ubiquitous perception will develop rapidly; the friendliness of equipment for human being will be greatly improved, and will have functions of self-calibration, self-diagnosis, self-healing, etc.; manufacturing systems will present the situation of man-machine collaborative work with human as decision-making body; and with the help of abundant information, more efficient production organization and scheduling will be formed, utilization rate of equipment, will be increased and production efficiency of manufacturing industry will be improved. In the development of the intelligent manufacturing system based on ubiquitous information, the intelligence level of equipment will be essentially enhanced,; the manufacturing mode and manufacturing means will no longer meet user's needs passively, they will actively perceive the changes of the user's scenarios, carry out information exchange, and provide active services according to the user's personalized needs of people. It is foreseeable that human—centered, personalized, intelligent and friendly ubiquitous information environment will become the inevitable trend of future information communication society and will bring the innovation service mode and service means for the modern manufacturing industry.

# 2 The Features of Information Manufacturing Systems Based on Ubiquitous Perception

As mentioned above, it is not difficult to see that information manufacturing based on the ubiquitous perception becomes a new historical stage of development of intelligent manufacturing system with human machine compatibility. Manufacturing industry supported by ubiquitous information technology will have obvious changes in the aspect of enterprise organization and processes, with the following features:

(1) Socialized service of manufacturing technology will become a trend. Supporting technology and service capabilities with relation to manufacturing will be upgraded greatly, the socialized resources and services demanded for manufacturing will emerge continuously and enriched gradually.

(2) Global production organizations without boundary will become the mainstream. Prefessional manufacturing services enterprise will run with high-efficiency, and full life cycle of the manufacturing process will be completed by diversified enterprises all over the world with the approach of seamlessly integration of socialization in order to really realize the boundaryless organization of manufacturing.

(3) Process optimization will be the regular business activities of enterprises. Process optimization is the main indicator elements of the core competitiveness of manufacturing enterprises, and will gradually transform from Business Process Reengineering to the business process structure of personalization and customization.

(4) The innovation operation mode for enterprise will become a diversified trend. With the development and integration of simulation and physical simulation technology, product design will approximate to the real product, "design means manufacturing" will bring sufficient imagination space for enterprises to innovate the operation mode.

With the support of ubiquitous information technology and systems, the new operation mode of enterprises has the following features and trends:

(1) The relationship among the manufacturing enterprises, the customers and the market turns closer. Because tracking capability of the manufacturing

process is greatly improved, the monitoring and services means of products after sale get enhanced, the management mode focused on production-centered with batch-manufacturing as target is transformed to the individuation manufacturing mode with customers' needs as main demand, simultaneously taking into account the customization of production efficiency, full life-cycle management.

(2) The reduction or flat of management-levels provided with basic conditions, decision-making become more effective. Flat management idea mainly pursues the requirements for management efficiency. With the support of ubiquitous information technology, socialized labor division of enterprises will be more professional, and each enterprise which completes the entire manufacturing process chain has the natural equal status. Flat management is not only the demand for management efficiency, but also really has the organizations prerequisite and supporting conditions of achieving flat management.

(3) Production organization is more flexible, and will adapt to the demand of market changes in the level of socialization integration. With the professional and high efficient running of enterprises and socialization seamless integration of manufacture resources, manufacturing industry can be reorganized in time in the socialization environment as boundaryless enterprises to achieve larger resources integration.

"Ubiquitous perceptual information manufacturing" is the direction for advanced manufacturing in the future. Looking back at history, it is not difficult to find that the information technology has pushed forward the progress of manufacturing industry, mainly in the fields of information acquisition, identification, processing, transmitting, retrieval, search, analysis and utilization, and then the new manufacturing mode such as digital manufacturing, and networking manufacturing and so on have come into being. Along this thought, in the next 50 years, the methods and means of information acquisition and identification technology ,such as RFID technology, and wireless sensor network technology, will be more and more abundant. Information processing technology will break through the traditional pattern of digital information processing, such as mode information processing, symbolic information processing for expressing and describing graphics and language. Channels of information transmission technology are more convenient, such as the ubiquitous network, communication satellites. On the premise of capacity of acquiring information greatly increasing, both the universality and hierarchy of information retrieval, analysis and use is extended and promoted. From the perspective of information technology progress, nowadays, the pursuit of digital manufacturing and network manufacturing can not represent all characteristics and main features of manufacturing in the future.

With the promotion of ubiquitous information technology, the changes and progress of the above-mentioned macro-level in manufacturing will have some new changes in the following 5 aspects that is five major function systems

   *Advanced Manufacturing Technology in China: A Roadmap to 2050*

of supporting advanced manufacturing:

## 1. Engineering Design System

With the development of computer and network technology, computer-aided engineering design system has greatly relieved the labor intensity of human, the computer's advantages in information processing has been given full play and is moving towards the direction of development of engineering design automation.

Engineering design automation refers to a technology of product design and analysis with the help of computer hardware and software as well as network environment support. It provides processes of product design, manufacture, assembly, analysis with computer support tools and means based on product data model. Engineering design automation not only runs through the entire process of product design and manufacturing, but also involves equipment layout, material flow process, production plan and cost analysis aspects of enterprises. Its application could serve to shorten product development cycles, reduce product development costs and achieve the purpose of optimal product design. Generally it includes computer aided design, computer aided process design, computer aided manufacturing, computer aided engineering, product data management systems and so on.

Computer aided design mainly went through the following stages: two-dimensional computer-aided drawing, three-dimensional product modeling, parametric design and feature modeling design stages which represent the future direction. Computer aided process planning is mainly divided into three types : derivation, generative and comprehension. Computer-aided manufacturing experienced different stages of development: manual programming, NC language programming, CAD/CAM direct programming, NC automatic programming. Computer Aided Engineering is a technology closely related to manufacturing field. Computer aided engineering for mechanical and electronic products design which are closely related to mechanical properties mainly benefits from the progress of finite element technology, while computer-aided engineering for other products is not developed to such a stage. For instance, products design of chemical products, and medicine associated with molecular reaction dynamics are still dependent on the experimental methods. The development of product data management mainly shows expanding application scope, the current product data management mainly focuses on document management, workflow and project management, product structure and configuration management as well as system integration.

It is predicted that by 2030, engineering design system will still based on digital technology, with engineering design automation as its main objective. On the premise of the progress of multiple fields digital modeling technology, it will realize engineering design automation system with stronger sense of immersion, knowledge fusion of multiple domain, multi-scale design (in particular, to extend to maximal scale or minimal scale), high integrated value, based on product data model.

It is expected that engineering design will transform to the virtual reality design on the basis of design automation from 2030 to 2050. Virtual reality of engineering design mainly refers to utilizing the new progress of information processing and information transmission technology, parallel adopting digital information processing, pattern information processing and semantic information processing technology, breaking through the limitation of digital information processing; using micro-sensor technology, expanding observation methods of product in microscopic scale in order to enhance the understanding of the physical objects and modeling standards; synthetically use multi-disciplinary knowledge and information processing technology such as physical, chemical and so on, overcoming lack of knowledge of material world in microscopic scale; realizing the collection of highly synchronous global professional knowledge from different places by utilizing ubiquitous information network technology; with the support of X-tronics technology of rapid assembly with various scales, integrating information model and rapid physical model of product to achieve hardware-in-the-loop (semi-physical) information aided products design and analysis technique highly consistent with the products.

## 2. Manufacturing Information System

Manufacturing system mainly consists of equipment subsystems, material transport and storage subsystems, manufacturing information subsystem and energy flow subsystem and so on. Of which, manufacturing information subsystem is the core of modern manufacturing automation systems, and the key of normal and optimizing operation of the whole manufacturing automation systems, including computer control systems, database management systems, networks communications systems, etc. It is primarily responsible for manufacturing information processing, logistics management, monitoring and control of manufacturing process, production planning and production scheduling.

The symbolic progress for Manufacturing Information Systems is manufacturing execution systems which starts in the 1980s, and got rapid development in 1990s. Manufacturing Execution System is the connecting bridge between program management layer and the bottom control, whose main functions include: resource allocation and status management, process-level detailed production planning, production scheduling management, document management, on-site data gathering, human resource management, production quality control, production process management, production equipment maintenance management, product tracking and product data management, production performance analysis and so on.

By 2030, with the rapid development of ubiquitous perception technology, the real-time tracking capabilities of material flow in the production line will be greatly improved in both time and space dimensions, the functions of

   *Advanced Manufacturing Technology in China: A Roadmap to 2050*

manufacturing information systems will be expanded widely, and the functions of function modules relevant with MES time and space will be upgraded essentially. Consequently massive information caused by these improvement will make the production process much clearer and transparent for the administrators with the push of information processing technology, the speed and quality of production decision making will is updated obvious. Inside the workshop, production will be expand toward the direction of the refinement control and quality forecasting of manufacturing process. Meanwhile, for the outside workshop, information management related to material flow (logistics), quality, planning will cause a chain impact, real-time information services will be provided for enterprises information management, and non-differentiation or plat of information management levels will be really realized. From 2030 to 2050, intelligent manufacturing system which develop people's creative capacity and has human intelligence will become the main theme. The future intelligent manufacturing system will show better characteristics associated with human intelligent behaviors, such as understanding language, learning ability, ability to logical reasoning and problem-solving, the mechanisms of human brain activity can be understand, the part of the brainwork of human.

## 3. Equipment Intelligent Systems

Manufacturing equipment is the main part of hardware of manufacturing automation, and mainly includes processing equipment, such as special automatic machine tools, combined machine tools, NC machine tools, machining centers, distributed digital control, flexible manufacturing cell, flexible manufacturing systems, flexible production lines, as well as measuring equipment, auxiliary equipment, clamp devices and so on.

The development of manufacturing equipment is closely related to the progress of materials, technologies and control techniques. Materials play an important role in the development of equipment technology, and the corresponding technology is the core competitive technique of manufacturing equipment. Control techniques in most high-end equipment play the role of effective control of the running process.

By 2030, with the fast development of ubiquitous network and ubiquitous perception technology, the cost of controls will be reduced. Therefore, on the one hand, the popularization of manufacturing equipment controlling application will be promoted; on the other hand, the application level of control will be greatly improved by comprehensive control application of multidimension, multidimensional rate, multi-scale information. Self-maintenance, self-recovery of equipment will become a reality in the health and safety maintenance of equipment, self-diagnosis of equipment. By 2050, intelligent level of equipment will be increased essentially. According to the changes of environment and task, equipment will have not only the ability to adapt parameters adjustment, but also the ability to adapt to structure. With the progress of materials and information technology, self-evolution and upgrade ability of equipment will be prompted, and the intelligent level of equipment will develop from the controllability and automation to the high intelligence stage of self- maintenance, self-adapting and self-evolution.

## 4. Enterprise Management Information System

Management Information System is a system composed of human and computer which can carry out collection, transmission, storage, processing and maintenance of management information. Enterprise management information system integrates the advanced management ideas and operation modes of enterprise into it, with enterprise as target.

The study of Management Information System started from the 1940's, during which experienced MRP of 60s, closed-loop MRP of 70s, MRP II of 80s and the ERP period of 90s. In recent years, ERP gradually began to be integrated into relevant content of CRM, APS, and business intelligence, etc.

By 2030, with the growing up of ubiquitous network and business intelligence technologies, ubiquitous information services for manufacturing will be achieved. Enterprise Information Management System from the current focus on digital information processing and services stage, step forward to ubiquitous services mode of the provision of business-oriented information processing separate back-end from front-end. In 2030 to 2050, various component parts of enterprise information management system will be achieved entirely free assembly, barrier-free and recombining of information systems, flexibility of information processing on the basis of more basic level and smaller particle size.

## 5. E-commerce Systems

E-commerce in broad sense refers to electronization of the whole business activities of advertising, trading, transaction, payment and service, with the help of electronic technology. E-commerce system oriented for advanced manufacturing is build for the manufacturing enterprises, traditional management ideas and concepts must be abandoned. With information technology, especially e-commerce technology,  the procurement, production, sales, and finance involved in all manufacturing enterprises alliance are integrated in an e-commerce platform. It helps not only to increase the directness and transparency of business, thereby improving efficiency and increasing business opportunities, to achieve innovation of manufacturing and business mode, but also to share resources and reduce costs, so that core competitiveness of the enterprise is improved and adaptability in complex and competitive environment of the enterprise is enhanced.

E-commerce experienced three development stages, i.e., e-commerce based on electronic data interchange EDI of 1970s to 1990s, e-commerce based on Internet of 1990s, and the e-commerce based on pervasive computing of the early 21st century.

By 2030, e-commerce based on ubiquitous computing will focus on

 *Advanced Manufacturing Technology in China: A Roadmap to 2050*

the advantages of mobile commerce, and is mainly embodied in the personal information management, mobile banking, mobile shopping and transactions, and mobile entertainment without the limitation of time and space to realize ubiquitous e-commerce functions. From 2030 to 2050, service function of e-commerce will be greatly upgraded, large-scale manufacturing and marketing will no longer exist and will be replaced by manufacturing on demand and individual marketing mode adapting to the dynamic nature and infinite subdivision of market, increasing diversity of user's needs, integration of ubiquitous manufacturing and business environment, integration of manufacturing and business process, resulting from the seamless collaboration trend of organizational structure. E-commerce activities can not only carry out in the virtual world with sound, light and information, and remote high-sync-perception network technology will realize the network transmission and reappearance (reproduction) of superficial sensory. It is almost possible for people to do business activities in the Quasi-reality network environments; e-commerce really enter into the stage of full true service.

Intelligent manufacturing system of human machine compatibility promoted by ubiquitous information technology, first of all is bound to realize intelligent manufacturing of human machine compatibility on the basis of promoting development and progress of a number of key technologies. The eight key technologies will be listed and discussed in the following of this book to some extent with the consideration of the distribution and importance of the key words.

In short, although intelligent manufacturing system of human machine compatibility is considered from the dimension of information development, which can not give the overall prospects of the development of future manufacturing systems, the whole development orientation of manufacturing systems and technology is discussed with a clear skeleton in this paper.

| | By2020 | By2030 | By2050 |
|---|---|---|---|
| | Establishing basic ubiquitous network,forming rudiments and typical application based on ubiquitous information manufacturing | Information manufacturing based on ubiquitous perception will get rapid development,pleasant level for the equipment will be greatly improved. | Establishing intelligent manufactuing systems based on ubiquitous information, achieving ubiquitous information services for manufacturing. |
| Engineering Design System | 3D design with parametric modeling as the core will have been widely used | Achieving engineering design with multi-domain knowledge integration, multi-scale, high integration level. | Transforming into optimum design of virtual reality |
| Manufacturing Information System | Achieving engineering design with multi-domain knowledge integration,multi-scale,high integration value. | Information management level will become undifferentiated and flat | Establishing Intelligent Manufacturing Information System |
| Equipmeng Intelligent System | Manufacturing epuipments develop to digital,high-precisional and flexible direction | Ubiquitous information promotes popularization and improvement of manufacturing equipment control application / Self-diagnosis,self-maintenance,etc of equipment will become a reality | Intelligent level of equipment will be improved essentially |
| Enterprises management information system | Pay much attention to information system focused on digital information proccssing and service-oriented | The emergence of ubiquitous service mode with the characteristic of mobile computing and networking applications | Realizing random integration and intelligent processing of information systems |
| E-commerce systems | E-commerce based on ubiquitous network | E-commerce based on Pervasive Computing represents advantages of mobile commerce | E-commerce enters into the stage of full-reality service |

# 3 Key Technologies



## 3.1 Ubiquitous Sensing Networks

### 1. Demands and Challenges

Manufacturing-oriented ubiquitous networks are the integration and fusion of wireless sensor networks, logistics management-oriented RFID networks, industrial control networks and enterprise information networks. It is an important mean of information interaction for human-human, human-machine and machine-machine in future manufacturing environment.

As a major future direction of manufacturing industry, information manufacturing systems based on ubiquitous sensing present new requirements for the development of manufacturing network technologies. In view of designing, factory site and information of equipments need Internet access,remote, and parallel design. In view of equipments, the sequential acquisition of wear and ageing states for equipments is needed, and potential problems and the optimal maintenance time can be predicted. In addition, trouble repairing time should be reduced, and the efficiency of equipments should be increased. The life of equipments also should be prolonged. In view of process control, full acquisition of process data and product quality information can provide the basis of optimal control strategy. In view of management, healthy state and running efficiency of equipments and position information oriented factory assets tracing should be acquired timely to promote automation level of factory management. In a word, ubiquitous network technologies should meet the requirements of intelligent manufacturing and provide plenty of information acquisition and transmission services in time and space.

For aforementioned developing demands, ubiquitous network

technologies have to face the following challenges:

(1) Ubiquitous interconnection and information transmission of large-scale sensor nodes include the integration of large scale wireless micro sensor networks with high reliability and low energy consumption and RFID integration.

(2) Information flow integration and remote access in enterprise interior include the wireless-oriented monitoring network for factory management layer, workshop monitoring layer and equipments layer on site, localization and tracking of low cost wireless technologies, and the fusion of wireless monitoring network and Internet.

## 2. State of the Art and Developing Trends

Up to now, ubiquitous network technologies oriented future intelligent manufacturing are still in the initial stage.

With the development of sensor technology based on MEMS technology, sensor networks will enter manufacturing field. In view of micro sensor networks with high reliability and low energy consumption, single-hop star topology is still the most reliable network topology used by present factory. As IEEE 802.15.4 supplements cluster tree network topology and IEEE 802.11 proposes Mesh structure network topology, multi-hop networks with center control emerge and multi-hop mesh structure also has become the research topic of factory wireless sensor networks. Due to the application of centralized control designed for wired network, the scale of present factory wireless sensor network can only support hundreds of nodes. Network performance is affected by the variation of factory environment greatly, since the power supply of these sensor nodes depends on battery. With the development of distributed intelligent self-organization technology, cognitive radio technology, environment energy acquisition technology, future factory wireless sensor network will accommodate thousands of nodes. As a result, the sensor nodes can acquire energy from environment and the network can adjust automatically according to the variation of environment for stable performance.

In terms of integration of wireless sensor networks and RFID, people begin to construct new information service system by using continuous physical information in time provided by wireless sensor networks and positioning mark information provided by RFID, for example, WISP in INTEL and temperature sensor tag in Bisa. Technologies provide efficient solution for processing and transportation of food sensitive to temperature. In order to integrate sensors and RFID information more effectively, researchers in UCLA and Sybase present middleware solution of REWINS (reconfigurable wireless Internet for networking sensor) and RFID anywhere respectively. Up to now wireless sensor networks and RFID technology have been developing respectively and their interaction is just in the initial stage. With the rapid increase of requirements for factory information service based on position in future, wireless sensor network and RFID should be completely integrated and fused in hardware and software.

In view of wireless-based industrial control network, low cost and short distance wireless technology represented by IEEE802.15.4 begins to enter equipments layer network. WirelessHART proposed by HART committee, WIA-PA proposed by China and ISA100 proposed by Association of the United States instrument are three main wireless network technologies. Profibus and other fieldbus standards have also begun to make wireless solution. For workshop, IEEE802.11 as a representation for wireless local area network technology emerges in workshop monitoring network. Cisco, Siemens and so on have issued products based on private protocols, and corresponding international standards are going to be published. For enterprise, people are considering to construct remote data exchanging network by using WiMax technology. At present, wireless technology is just a supplement of wired network technology. However, wireless technology will take replace of wired network technology once reliability, capacity, energy consumption and security of wireless technologies are resolved.

For localization and tracking based on low cost wireless technology, GPS systems with high localization precision are being used in factory now. However, we cannot use GPS in door or some special situations due to its costliness, high energy consumption and poor scalability. In addition, phone localization as a new location method has also been used in factory environment. Real-time localization and tracking service of Wi-Fi, ultra broadband and RFID technology have been adopted in civil use, and will be utilized in factory network. In view of localization technology, triangular and multi-point localization technology based on range are the mainstream technologies at present, while centroid localization technology with non-range is increasingly developing. The costs, precision and real time capability in present localization and tracking technology cannot be completely guaranteed. With the popularization and application in factory of wireless sensor network and RFID technology and the development of low cos assistant sensor ranging technology, low cost wireless network localization will become the leading localization technology for future factory and impel the development of factory positioning service system .

In terms of integration of factory wireless monitoring network and Internet, now factory monitoring network mainly connects with ERP system to provide equipments, process and logistics information for ERP system. Monitoring and control network only covers enterprise interior and Internet only can extend to factory ERP system. This leads into that real-time equipment information and sensor information in factory cannot be accessed remotely. To solve this problem, people are doing research on the embedded IPv6 network technology and IETF are making corresponding standards. Network layer protocol of factory wireless monitoring network will uniformly use IPv6 standard so that the remote manufacturing and management based on Internet may be possible in future.

## 3. Objectives and Tasks

Concerning on information acquisition and interaction in manufacturing process, interconnection and information transmission of ubiquitous large-scale sensor nodes and information flow integration and remote access in enterprise interior are two core research topics that need to be settled.Five kinds of key technologies need to be broken through, namely the networking technology of large-scale micro sensor with high reliability and low energy consumption, the integration technology of factory wireless sensor networks and RFID, the wireless-oriented monitoring network technology for factory management layer, workshop monitoring layer and equipments layer, the localization and tracking wireless technology with low cost, the fusion technology of wireless monitoring networks and Internet. Ubiquitous network technology system servicing intelligent manufacturing should be established to provide ubiquitous network services for design, equipments, process, management and business.

## 4. Developing Roadmap

Small-scale wireless sensor networks and RFID technologies for logistics management and predicted maintenance of equipments will be widely used in factory by 2020. Internet will be used to connect enterprise layer network, control layer network and equipment layer network, thus the remotely real-time access to the production and equipment information in factory will be realized.

In terms of interconnection and information transmission of omnipresent large-scale sensor nodes, the limitation of topology structure will be solved and the rapid networking of micro sensors by using same communication medium will be achieved. Dynamic maintenance, high cure capability and self-adaptability of network should be achieved. The same micro sensor should possess varieties of network protocols, which can enable it to access different networks according to different requirements. Integration middleware technology of wireless sensor network and RFID network should be broken through and inter-operation of wireless sensor network and RFID network should be realized. The oriented service system architecture will be established. The core equipment with the integration of reading function of RFID, wireless sensing and networking should be developed. As far as information flow integration and remote access in enterprise interior are concerned, wireless network technology for factory automation and remote wireless network technology in enterprise interior should be broken through and international IEC standard should be made. Interconnection and inter-operation of wireless network and existing factory wired control network should be realized. Factory wireless monitoring network can be widely used in monitoring, log and open loop control. Precision error of short wireless ranging technology is controlled within one in a hundred thousand by using accurate wireless ranging technology. In order to apply accurate ranging in factory widely, the costs of accurate ranging sensor should be reduced. Multi-target tracking should be realized and real time prediction for targets' position can be achieved with history knowledge. Wireless monitoring network of factory should be connected by Internet, and monitoring data of factory should be acquired timely. Distributed manufacturing should be realized. The monitoring networks for different factories work together to finish particular manufacturing assignment and collaborate to adapt to different requirements.

Up to 2030, large-scale wireless sensors and RFID mixed network will be extensively used in factory. The promotion of wireless factory control network will make manufacturing system more flexible and remote design of manufacturing system possible.

In terms of interconnection and information transmission of omnipresent large-scale sensor nodes, the distribution and acquisition of communication resource for no-center network should be solved and distributed ad hoc network should be realized. The equipments should adjust automatically their roles in network, and choose work method according to time and space information. The environment perception and adaption technology of network should be broken through. RFID network and wireless sensor network should be integrated in terms of protocol architecture, and entities (label, reader) in RFID network should carry out direct communicating cooperation with wireless sensor nodes. RFID labels should be equipped with the basic function of wireless sensor network nodes. Precise assets localization and tracking systems for factories should be developed. In terms of information flow integration and remote access in enterprise interior, the coexistence of industry wireless networks with different standards and the coexistence between industry wireless network and varieties of business wireless network should be solved. High security technologies and high usability technologies for industry wireless networks should be conquered. Factory wireless monitoring networks will be widely used in low speed process control and partly replace low speed wired control network. Data fusion technologies of large-scale network should be broken through and targets should be localized rapidly according to ranging values of massive sensors. The limitation of real-time localization speed should be solved, and the timely tracking for moving target with high speed should be achieved. The problem of anti-interference for wireless localization should be figured out, and the performance of localization and tracking for targets in severe environment should be promoted. Global Internet of things can be formed by using IPv6, and products under monitoring in full life cycle should be implemented by monitoring system with Internet. At the same time, usable states and customers' requirements should be fed back to monitoring system of factory by Internet, and reconfigurable factory control system should rapidly adjust manufacturing process to produce the products by resetting, reusing and machine's reprogramming automatically, which are required by customers and markets.

By 2050, Plug and play sensors with RFID function should be developed in the range of factories, and should connect Internet automatically. Wireless-based factory control network will be achieved then. Autonomous reconfiguration of manufacturing systems according to the variation of environments, resources and assignments can be implemented.

In terms of interconnection and information transmission of ubiquitous large-scale sensor nodes, the limitation of different network transmission medium should be broken through. Heterogeneous micro controller based on interconnection technology, and the inter-connection and inter-operation of heterogeneous network should be realized. Connection whenever and wherever, adaptive topology, adaptive networking and network maintenance should be realized. RFID with very low cost should have fast reading capability with long distance, and wireless sensor nodes can also acquire energy at the same time. These two technologies can unify in hardware and software. In terms of integration of information flow and remote access in enterprise interior, broadband spread spectrum technology and time, space and frequency optimization sharing technology of channel resource will be broken through to reach 50Mbps throughput for large-scale factory wireless monitoring network. Energy acquisition and transmission technology in factory environment should be broken through. Factory wireless monitoring network with high speed close loop control and security control will be widely used. Factory control systems with complete wireless technologies will energy. Inter-localization and tracking between different massive dynamic targets should be realized to provide ubiquitous target localization and tracking service. Collaborative utilization and uniform decision-making for varieties of localization and tracking technologies should be realized to supply target localization and tracking with high real-time capability, high precision and low cost. The fusion between factory monitoring network and Internet will form logistics and cooperative manufacturing with globalization and individualized customers-oriented manufacturing system.

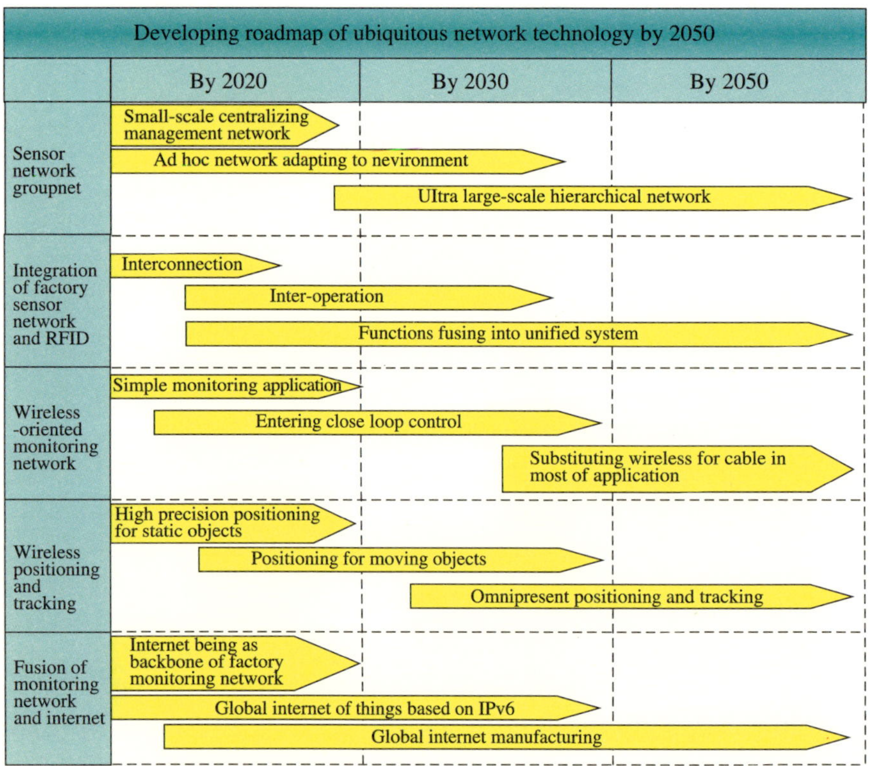

### 5. Summary

Ubiquitous network technology is the fundament technology that realizes human-machine harmonious intelligent manufacturing. The core of ubiquitous network technology is acquisition, transmission and the integration of signal generated by omnipresent sensors. According to the developing roadmap in this section, two basic scientific difficulties will be solved and five key technologies in ubiquitous perception field will be broken through in three phases. This will provide ubiquitous network services for the new generation of future intelligent manufacturing.

## 3.2  Ubiquitous Manufacturing Information Process

### 1.  Demands and Challenges

With the development of wireless network technologies, wireless RF identification technologies, information technologies and cognitive science, the performance of acquisition, identification, processing, transmission, index, analysis and utilization for information will be promoted. People's capability in controlling and using information is expanded greatly. Ubiquitous information manufacturing drove by perception favors the advent of new generation of information manufacturing and automation. Ubiquitous information manufacturing will definitely be a hot topic of information processing technology in future manufacturing field.

Environment constructed by ubiquitous information manufacturing is to naturally and profoundly accommodate varieties of terminals with environmental perception capability, mobile communication, information acquisition, context awareness, intelligent software, human-machine interaction into every field in manufacturing industry. As a result, more precise, more agile, more intelligent, more personal and friendlier human-center manufacturing environment will be formed. The objective is to establish a human-machine harmonious environment that is full of calculation and communication capability. In this environment, people can transparently acquire digital service anytime anywhere. The objects of information acquisition and processing not only concern structured data, such as process parameter, equipments status and business process information, but also realize high centralized and fused with multimedia information, such as sound, image, graph and text. The seamless joint of multi-dimensional perception information between physical manufacturing space and information space can be achieved. This will efficiently instruct manufacturing process of actual world. Huge challenges and urgent requirements for ubiquitous information process is how to unify and make full use of these information resources, and indicate high precision, high intelligence,high personal information expression and process method in the new generation of manufacturing process.

## 2 . State of the Art and Developing Trends

At present, all research on ubiquitous information process are mainly focused on ubiquitous computing, ubiquitous communication, ubiquitous perception and human-machine interaction and so on. In the 1990s, Mark Weiser firstly proposed ubiquitous computing and proposed information society morphology of computers or terminal equipments that could communicate with network. In the 2000s, plenty of corresponding element technologies emerged, such as nano technology, biochip technology, micro system technology, wireless communication technology (Bluetooth, Wi-Fi, UWB, Zigbee, etc.), cognitive engineering technology and flush type technology. Meanwhile, new ubiquitous information process network, such as BAN and PAN, was also proposed. These technologies accelerated the development of ubiquitous information process service , which is the core of intelligent interconnection, resource sharing and collaboration service. A large number of research and roadmaps on ubiquitous information process has also begun, such as European FP6 plan, American general computing and Japanese U-Japan plan and so on.

With the development of aforementioned element technologies, corresponding research for varieties of application filed is starting. Especially in manufacturing field, as the products' performance is improved and its structure is becoming more complex and fine, as well as the functions are becoming various, information flux of product line, product equipments interior, manufacturing process and management assignment will increase sharply. This drives the transition from traditional manufacturing to agile manufacturing, networked manufacturing and green manufacturing. The utilization of ubiquitous information technology enriches the method of acquisition and identification of manufacturing information, such as RFID technology and wireless sensor network technology. The method of information transmission, such as Internet, wireless networks, satellite and sensors, is more convenient., The acquisition, identification, processing, transmission, index, analysis and utilization of information will change from traditional digital information processing  (such as expression, diagram, language and symbol) to fusion and processing of multi-mode information. This is beneficial to improving the ability, efficiency and scale for modern manufacturing system. America, Japan and Europe all proposed related research projects. China also proposed similar research projects. Generally speaking, information manufacturing and ubiquitous information processing are in the stage of conception and experiment, However, all the above countries have listed ubiquitous information manufacturing as the future country developing plan, and thus it will be doomed to be the future direction of advanced manufacturing.

## 3. Objectives and tasks

Manufacturing industry should make full use of the interaction advantage

   *Advanced Manufacturing Technology in China: A Roadmap to 2050*

of manufacturing science, information science, life science and physical science in future development. More intelligent, smarter and more personal ubiquitous information processing platform should be constructed by information acquisition, perception, reasoning and decision-making. The perceptive ability and efficiency of manufacturing process for massive, unstructured and hierarchical perception information should be improved and the difficulty of processing massive information in manufacturing process should be solved. New manufacturing mode based on ubiquitous information processing will be established. The development of manufacturing and information industry will be impelled and the whole capability and level of manufacturing industry will be promoted. In order to fulfill aforementioned objectives, we have to conquer the following theories and key technologies .

**1) Fundamental theory framework of synthetically ubiquitous information processing for manufacturing process should be founded.**

(1) Multi-mode ubiquitous perception information integration and semantic modular technology for manufacturing process. Efficient computing model for ubiquitous sensing collaborative information computing model for ubiquitous perception information and complex spatio-temporal event semantic model for manufacturing process.

(2) Cognitive mechanism and human-machine interaction modeling technology based on the fusion of sensing information. Context-aware and reasoning for human-human, human-machine and machine-machine, semantic information index for massive perception information, the design of more personal, friendlier, easily operated and intelligent human-machine interface.

(3) Multi-mode identification technology and validation system based on the fusion of sensing information.

**2) Key technologies framework of ubiquitous information processing for manufacturing process should be founded.**

(1) environment-aware modeling technology for manufacturing.

(2) structured/unstructured trans-mode and trans-scale ubiquitous information integration technology.

(3) massive sensing information processing technology based on cloud computing mode.

(4) wearable computing and contex computing technology in ubiquitous information manufacturing mode.

(5) complex Event Detection technology under spatio-temporal dynamic mode.

(6) context-aware middleware technology for ubiquitous information.

(7) discovery technology for manufacturing resource services.

(8) rich Internet applications programming technology.

**3) Synthetically ubiquitous information processing platform system for manufacturing process should be founded.**

Ubiquitous information processing platform for manufacturing process should be developed so that the harmonious human-machine interaction platform environment can be formed. Massive sensing information is processed with the retractile performance, intelligence and high efficiency. Combination with knowledge engineering method and system engineering method, knowledge database, model database, method database and multimedia database of sensing information for manufacturing process should be designed. Intelligent science management method, such as intelligent prediction, intelligent planning, intelligent optimization and intelligent decision-making, will be introduced into the platform to promote its whole intelligent performance. Intelligent and friendly multimedia human-machine intelligent interface should be designed. Tool database, component database and management model should also be designed.

## 4. Developing Roadmap

According to the research trends all over the world, future manufacturing industry will definitely adopt new technologies and new modes, while ubiquitous information processing that will integrate ubiquitous network, ubiquitous perception, ubiquitous service and ubiquitous intelligence will become the representative new technology and mode in future manufacturing industry. Taking into account objectives in the following 50 years, the development of manufacturing industry with ubiquitous information technology as the main method will undergo a stepwise process. The process not only relates to interdisciplinary science research, but also the development and establishment of ubiquitous information processing service platform with high performance for manufacturing field. At the same time, the realization of demonstration and validation for typical manufacturing industry and achievement transform are also concerned. The synthesis solutions for the development of future manufacturing industry will be achieved finally.

From 2010 to 2020, ubiquitous network infrastructure in manufacturing environment should be constructed and improved gradually. Taking the infrastructure as the basis, intelligent processing method for synthetic perception information and the research on collaborative computing model in manufacturing environment can be carried out. With the increasing application, the requirement for massive information, manufacturing information and perception information needs to be integrated, and the integrated process for multi-information is also demanded. The related technology processing system should be established, and technology reconstruction and innovation for equipments, techniques, environment, energy sources and business flow in traditional manufacturing process should be finished.

(1) Transition from traditional manufacturing unit to ubiquitous sensing based intelligent manufacturing Traditional manufacturing unit relates to equipments, techniques, environment, resources, state and energy sources power and so on, and these unit technologies will create new intelligent

   *Advanced Manufacturing Technology in China: A Roadmap to 2050*

manufacturing units by using micro electromechanical technology, flush type technology and sensor technology. New manufacturing process will be aware of information wherever, and based on these information, intelligent prediction, intelligent planning, intelligent optimization and intelligent decision-making will be achieved.

(2) Acquisition and fusion for sensing information of massive manufacturing unit and manufacturing process information. The combination of sensing information of manufacturing unit and specific business information, by the integration of equipments context, business context and process context will enable manufacturing process under control every aspect of manufacturing and automatically make decision to realize more exact and synthetic manufacturing process management. Meanwhile, the acquisition and fusion of massive information is the biggest challenge in this stage.

(3) Awareness and reasoning based on sensing information. Future manufacturing process will be mostly the collaborative manufacturing at different times and spaces. Sensing information integration and awareness reasoning based on multi-source hierarchical play an important role in collaborative manufacturing process. Cognition in-depth based on sensing information will inspire a variety of new methods for production scheduling.

From 2020 to 2030, based on the improved ubiquitous network, the general performance of ubiquitous sensing and ubiquitous service should be enhanced further, such as tracking environment in manufacturing process, adaptive performance for varied conditions, self-learning ability for uncertain issue in manufacturing process, self-planning and self-organizing capability for manufacturing process based on perception information. Moreover, human-machine interface should be improved, and the semantic understanding and fusion for perception information should be automatically finished. In tran-spatio-temporal manufacturing environment, the integration of manufacturing resources management, context-aware of manufacturing process, and complex event of manufacturing process will be broken through.

(1) Resources management of tran-spatio-temporal manufacturing process. In the tran-spatio-temporal manufacturing environment, the variations of manufacturing resources not only relate to hard resources, such as equipments, techniques, flow, and personnel, but also soft resources, such as plan, service, management and so on. Adaptively fulfilling registration, searching, discovery, selection, matching, optimization, evaluation and assembly of resources service can realize more agile sharing and collaboration for manufacturing resources and hence carry out effective manufacturing resources service and management.

(2) Context awareness of tran-spatio-temporal manufacturing process. The tran-spatio-temporal manufacturing process will synthetically deal with all kinds of context information including time, space, varieties of technical parameters, workpieces, and equipment state and so on. The sensing of manufacturing information for different context plays a key role

in collaboration of multi-scene manufacturing process. The application of context-aware middleware for manufacturing process can process multi-scene collaborative manufacturing and effectively solve inter-operation problem caused by heterogeneity between equipments and software system in ubiquitous network environment. This will promote the flexible application customization for context-aware.

(3) Complex events integration of tran-spatio-temporal manufacturing process. Massive events caused by massive sensing information indicate complex information of equipments, techniques, processes, personnel in multi-scene tran-spatio-temporal manufacturing process. Traditional technologies are difficult to find significant knowledge in the massive information. However, the semantic modeling for complex spatio-temporal events in manufacturing process, seeking significant knowledge, and analyzing and exploring the cognition mechanism can further improve intelligent management performance in manufacturing process, such as intelligent prediction, intelligent planning, intelligent optimization and intelligent decision-making.

From 2030 to 2050, After ubiquitous networks, ubiquitous sensing, ubiquitous services and relating intelligent information processing have been widely used in manufacturing process, technologies characterized by ubiquitous intelligence will be applied in this period. Intelligence is not only indicated in production equipment, production setup, industrial sensor, RFID systems and so on, but also embodied in multimedia information, acquisition and integration for structured/unstructured information. At the same time, ubiquitous information processing service platform with high performance will be developed and improved. Varieties of intelligent computing modes, e.g. wearable computing and affective computing, will be openly integrated to establish harmonious human-machine interaction. The performance of self-configuration, self-recovering, self-optimizing, and adaptive of human-human, human-machine and machine-machine in the process of design, manufacturing and management process should be improved significantly. Overall research capability in this field should be enhanced greatly in order to provide new technologies for exploring and promoting the performance of control and management in manufacturing process.

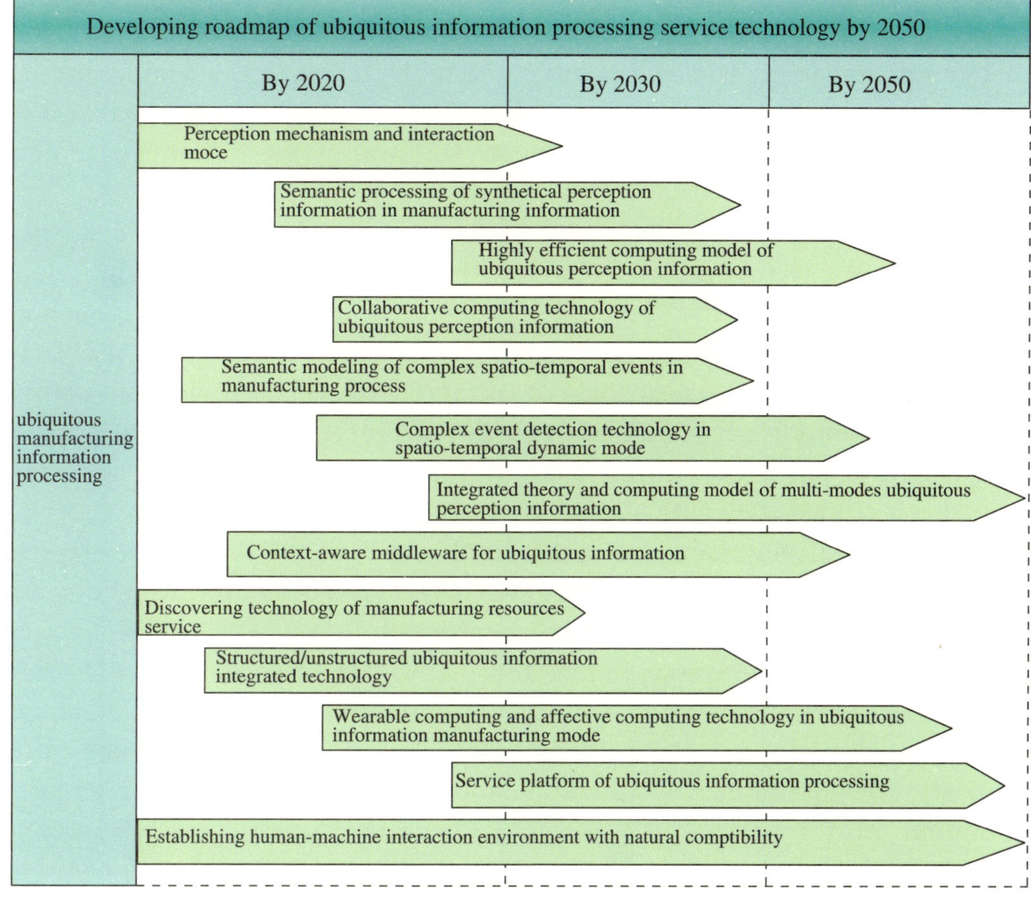



## 5. Summary

With continuous efforts, ubiquitous information processing will bring innovative energy and power for the future manufacturing industry. Intelligent terminals, mobile communication, information acquisition, context awareness, intelligent software and human-machine interaction technologies will naturally integrate into every aspect of manufacturing industry. The more precise, more agile, more intelligent, more personal and friendly manufacturing environment will be constructed, and thus the intelligent manufacturing era based on ubiquitous information is approaching.

# 3.3  Virtual Reality

## 1.  Demands and Challenges

Increasing numbers of complex and variable parts are involved in the design, manufacturing and assembly of key equipments. These parts have certain special characters, such as long developing cycle, slow updating processes, and often of high technical value-added. Hence, new technologies that could change the traditional design, manufacturing and assembly modes are needed by advanced manufacturing in the future. Clearly, virtual reality and real three-dimensional (3D) display technology are among these new technologies.

Taking computer technology as its core, combined with related science and technology, virtual reality technology can generate a digital environment which is similar to a certain real environment in some feeling aspects such as vision, hearing and tactility. Then, users can interact with those virtual objects in digital environment with the help of specific devices. The augmented reality, developed from virtual reality technology, can integrate the virtual objects generated by computer with their counterparts in the real world, and then construct a virtual space with virtual-real integration.

Real 3D display means that the depth effect of the displayed image is consistent with its counterpart in solid 3D real world which we can seen. The present 3D display card is a display card with "solid 3D image from two-dimensional plane projection". In fact, this method uses two-dimensional plane projection to create "look-like" 3D image illusion, However the projected image is still two-dimensional display essentially. Real 3D display technology is a kind of multi-disciplinary synthetic technology, which relates to electronics, computer, optics, mechanics and other disciplines. Its display property is suitable for analyzing 3D science, scene simulation and collaborative operation and so on. Hence, real 3D display will become the new technological approach to present virtual reality and augmented reality technology.

Virtual reality technology has vast application prospects in the complex equipments and product development fields, such as key equipment design, manned spacecraft trial, airplane design, ship design, and automobile design and so on. Hence, virtual reality technology has already attracted great attention in China and around the world. For example, it has been selected as one of the 22 leading technologies in Chinese middle or long-term science and technology development plan.

It is difficult for operators to determine the position relationships of different parts rapidly and precisely using the traditional two-dimensional display. This obstructs the further increase of the operator's productivity. Compared with the traditional two-dimensional display, the operators are more familiar with the real display of those parts in 3D world. Real 3D display (Figure 3.1) allows users to watch and walk freely with a large range of view field and distance, and fully reproduces the actual observation way of human beings on real scene. Hence, it can provide an effective technical method for the validation of 3D manufacturing tasks of key equipments, and greatly decrease the perception load of 3D manufacturing tasks. Moreover, the operators can acquire the information needed in their works rapidly and precisely, reduce the possibility of misestimating and improve their efficiency.

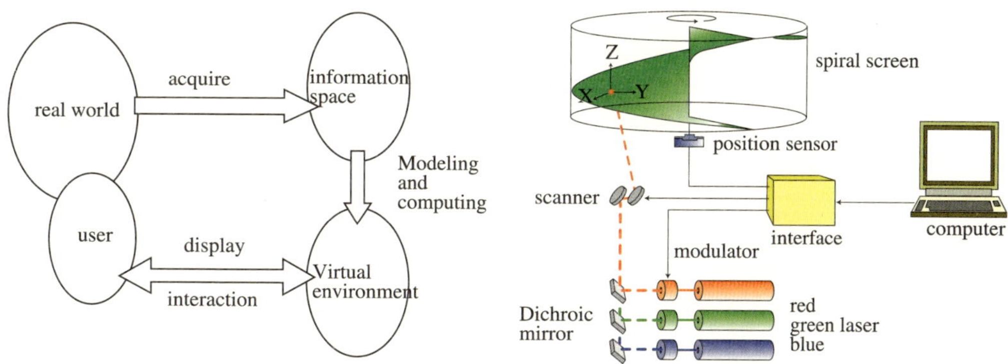



Figure 3.1   Virtual reality and real 3D display

## 2.  The State of the Art and Development Trends

A large number of major projects have begun to research and use virtual environments. Related to many engineering fields and containing many scientific theories and technologies, virtual reality technology plays an important role in the development of virtual environments because it can provide a 3D, emotional, and interactive supporting platform. Constructing an effective virtual environment based on simulation, collaboration and visualization technologies can decrease the cost of product development, shorten the design cycles and reduce the repeated engineering effort.

### (1) Virtual Reality Technology

Virtual reality is a kind of computer system which can create and experience virtual world. It fully utilizes the integration technology of computer hardware and software resources to provide real-time 3D virtual environments. Users can enter an virtual environment to watch the virtual world and hear vivid sounds generated by computer. Users can have their real feeling when they inter-operate with and even talk in virtual worlds.

Recently, collaborative working environment with virtual-real integration has been applied to virtual training, complex product design and maintenance, etc. With the applications of virtual reality technology industrial training and equipment maintenance process, when employees are staying beside the equipments and tools, and are operating the equipments, they can get visual guidance information so as to provide them correction and help in real time. There is increasing number of companies around the world using virtual reality technology. For example, U.S. Johnson Space Center uses Hubble telescope maintenance system; Boeing Company uses CATIA system for 3D simulation of Boeing 777's design and maintenance; European Computer-Industry Research Center (ECRC) utilizes virtual-real integration technology to explain the objects or the environment; Sony Corporation in Japan developed a virtual-real integration virtual prototype called Trans Vision; Volkswagen Automobile Co., Ltd. and Heinz Nixdorf Institute in Germany developed a vehicle ergonomics test platform with virtual-real integration technology in 2005.

With the development of virtual reality technology, virtual-real integration, portable interaction, and multi-channel perception will become the key technological needs of different application systems. Real 3D display technology and multimedia technology are the important components in virtual reality applications.

### (2) 3D Industry Worldwide

With the advance of industrial mass production, USA, Japan and other developed countries have firstly applied 3D display technology to those production lines mainly for executing dangerous and heavy tasks, such as intelligent assembly. Those 3D modeling and design software, such as Solid Works and Pro/E developed in these countries, have been successfully applied to large-scale industrial sites for digital modeling of products, 3D design optimization and analysis, strength/stiffness/stress analysis of parts, simulation and developing of digital prototypes, etc. and for functions such as parametric design, solid 3D modeling and complex system modeling. They have replaced the traditional flat drawings and increased efficiency. The well-known airplane manufacturing company, the Boeing Company, is directly using 3D paperless model to carry out the integrated design of its airplanes in Boeing brand family today. As shown in Figure 3.2 and Figure 3.3, it has gradually become a trend to replace the traditional flat drawings with 3D modeling in major manufacturing industries, such as airplane, automobile, and ship building.

Figure 3.2　3D simulated design for Boeing 777

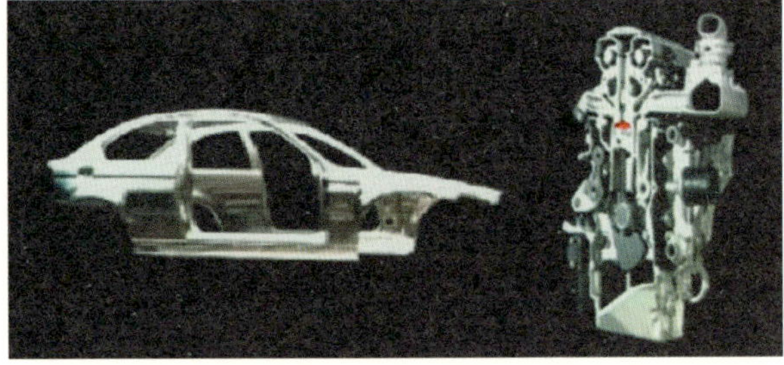

Figure 3.3　3D display and assembly

### (3) 3D Industry in China

Compared with the rapid development of intelligent 3D industry in foreign countries, 3D industry in China has not been applied as an emerging manufacturing philosophy to the actual production processes. This is mainly due to the short history of manufacturing industry in China, as well as the constraint of traditional industrial philosophy and principles. There are only individual enterprises that use 3D design system for product design and development. For example, Xi'an Aircraft Institute has developed full-scale airplane digital prototype to realize 3D design, 3D collaboration, and 3D pre-assembly. At present, only a small amount of large-scale enterprises have begun to use 3D design system for product design and development. In contrast, the majority of Chinese Small and Medium Enterprises (SME) still use 2D flat drawing as the standard for product parts' manufacturing.

In the future, Chinese domestic 3D industry will make major progress mainly in the design and manufacturing process simulation of 3D digital product, real 3D display, and their assembly and maintenance.

## 3. Objectives and Tasks

As an emerging and important research direction in information technology field, virtual reality technology has its significant application value in manufacturing industry. In recent years, some developed countries and regions have invested a huge amount of capital to keep their leading position in advanced manufacturing field. Therefore, they have accumulated plenty of technological achievements. To be faced with the future development trend of advanced manufacturing in the next 10 years, 30 years and 50 years, we must make our own technical development plan of virtual reality in the carefully selected areas, in order to assist Chinese manufacturing industry to achieve its leap-forward development and breakthrough in the future. Therefore, it is necessary to propose the objectives and tasks of those key technologies for future 3D industrial production.

First of all, China needs to learn the advanced philosophy of 3D manufacturing from foreign countries to further shorten product design cycles, reduce R&D cost and increase productivity. Note that very few foreign enterprises can use real 3D display and human-machine natural interaction mode to perform industrial production (Their process of 3D-oriented industrial production concentrates mainly on modeling, rather than human-machine natural interaction mode). Therefore, China can take this as a breakthrough point to further improve manufacturing efficiency. In order to accomplish these objectives, the following four main tasks need to be accomplished incrementally.

**Objective 1: Acquisition and modeling of 3D information (Design)**
- 3D drawing
- Part feature
- 3D modeling

This objective is to use 3D paperless model to replace traditional 2D flat drawing. As it is the basis of fully applying the final real 3D display technology to industry, it should be realized as soon as possible. Currently, some technologies in this field have become mature in foreign countries, and a lot of products with good performance, high degree of commercialization are emerging, and a wide range of utilization are emerging. China could learn a lot of experiences from them.

Figure 3.4    Acquisition and modeling of 3D information

### Objective 2: Position computing and space relation understanding (Manufacturing)

- Euclidean space calculation
- 3D Boolean calculation

This objective is to take 3D paperless model as the basis of determining the spatial position relation of all assembled parts. At the moment, estimation is still based on 2D media in China and foreign countries. However, this method is not consistent with human's natural habit of observing 3D objects. Thus, it can result in manual errors easily. The application of real 3D display can increase judgment efficiency and reduce manual errors. We can attempt to achieve this objective after the fulfillment of Objective 1.

### Objective 3: Construction of 3D behavior performing systems (Assembly)

- Complex assembly simulation
- Spatial motion simulation
- Virtual site synchronous performing

This objective is to fully apply the intelligent 3D industry to the entire industrial production processes. The information, which used to be perceived only when people go to the manufacturing factory site in person, can be obtained now from simulation models made up of hardware and software. After the above two objectives are achieved, this objective should be achieved to greatly increase the entire productivity.

Objective 4: Establishment of 3D display and human-machine interaction platforms (Environment)

- 3D space positioning with high precision
- Virtual-real integration with high precision
- Natural and compatible 3D interaction

This objective is made according to the characteristics of complex manufacturing systems, that is, they need technologies to acquire space position with high precision. Its purpose is to support the virtual-real model integration and various operations in those environments performed by participants. As far as dynamic environments of real 3D display are concerned, the integrated degree between virtual-real model and objects plays a key role in constructing 3D interaction environments (platforms). Most of the existing systems adopt relatively simple virtual-real model. However, in the future, 3D models considering shelter, light and other factors are needed to solve problems with high complexity. Seamless and real-time 3D display of virtual-real objects, natural and compatible human-machine interaction should be realized.

## 4. Development Stages

According to the objectives of intelligent 3D industry, the application development of 3D industry can be divided into three stages, each stage with its related focuses and research contents.

### Stage 1: From 2009 to 2020

Acquisition and modeling technology of 3D information will be widely applied to manufacturing sectors. It is expected that the penetration ratio of acquisition and modeling technology of 3D information in large and medium state-owned manufacturing enterprises will be more than 85%. Meanwhile, the technical results of industrial modeling with 3D display will be improved and applied steadily.

### Stage 2: From 2020 to 2030

Industrial modeling technology with 3D display will be widely used in manufacturing industry. It is expected that the penetration ratio of 3D display application in large and medium state-owned manufacturing enterprises will be more than 90%. Meanwhile, the research achievements of 3D scene computing and real-time drawing will be accumulated. Moreover, the achievements of simulation industry with 3D display will be improved and applied steadily to manufacturing fields.

### Stage 3: From 2030 to 2050

Research results of virtual site assembly with 3D display will be accumulated. Additionally, 3D display and human-machine interaction platforms (environments) will be built up and applied steadily to the actual manufacturing industry.

| Developing roadmap of virtual reality and real 3D display technology by 2050 | | |
|---|---|---|
| | **By 2020** | **By 2030** | **By 2050** |

**Virtual reality and real 3D display**
- Acquisition and modeling of 3D information
- Industrial modeling based on 3D display
- 3D scene computing and real-time drawing
- Simulated industry manufacturing based on 3D display
- Virtual site assembly based on 3D display
- 3D display and hurman-machine natural interaction platform and application

## 5. Summary

Based on the development trend in the next ten years, thirty years, and fifty years and the requirements of advanced manufacturing on virtual reality and real 3D display key technologies, the development roadmap here is proposed from the perspective of future 3D industrial production. Hopefully, it can assist the leap-forward development and breakthrough of Chinese manufacturing industry in the future. Finally, the figure below is introduced to present the main contents, requirements and objectives of future 3D industry.

# 3.4 Human-machine Interaction

## 1. Demands and Challenges

The interaction between human and manufacturing equipment has always been the key issue to determine the manufacturing level. When human being used only simple production tools, the relationship among human, production process and production tools was simple. Manufacturers were in charge of tools and production. When human being gradually invented more and more intelligent machines, they began to use power and created tremendous productivity, and the complex relationship of human and production equipment was emerging. The invention of machine have shorten the distance of human and objects produced and hence human have to propose various method to operate and control increasingly complex machines and manage increasingly large production systems. At the same time, human have to work with machines together. Human-machine interaction has always been a research topic. With the development of advanced manufacturing technology, products diversification is required by markets. This impels human being to exert their intelligence positively. Human-machine interaction faces new challenges and developing opportunities.

## 2. The State of the Art and Developing Trends

Since human began to use machines in manufacturing process, operation and control system of machine was emerging. Human communicate with machines by using operation and control system. The operation and control system can be very simple, for example human observe instrument or directly observe machining state, and use handle to manipulate the production equipment, while the present operation and control systems basically adopt automatic control system. On the other hand, since the machines appear in the manufacturing process, human have to cooperate with machines to finish work tasks. The production equipment is also far from full automation.

The operation and control of human to machines can be characterized by using pre-compiling programs for machines making system to finish predetermined objectives. In this process, the automatic control system obtains target information by measuring temperature, pressure, flux (in flow industry) or tool geometry coordinates and path, etc. A variety of control algorithms are used to adjust system behavior automatically to achieve the predetermined objectives.

However, the operation and control of human to production equipments is limited. Take high-precision machine tools for example, efficiency of machine is crucial for high efficiency and cost-effective production, but debugging before utilization is time-consuming. We can not precisely predict when machines can reach its working state because both machines' own temperature and ambient temperature affect the performance of machines. It is important to guarantee the machining quality, and the deviations caused by abrsion after long tome running of machine tools is unbearable. The current approach is estimate. It is clear that, the operation and control of human to machine tools is limited by inadequate information.

Robots, which can flexibly manufacture machines, satisfy the different tasks and possess the capabilities of conveniently, rapidly changing work position and object in time. Unfortunately, in order to meet these requirements, the existing robots have to spend large amount of time and labor. A major reason for robots' flexibility is that it can adapt to various tasks by changing procedures. The problem is programming, but the programming is a bottleneck. The current robots' programming is mainly dependent on manual teaching, namely users use the language in teaching box (Control Panel) to instruct the robots how to walk and how to do step by step. When the task is changed, the time-consuming work needs to be done again. Automated programming can not be widely used, because the deviation between the calculated position and actual position which the robots' end effectors (such as arc welding pliers) want to reach is not known precisely. If the position has deviation, people have to adjust it. This enables flexible robot to be applied to the occasion of mass production. However, for these fields which are in urgent need to use robot as the main force such as airplane assembly, they are helpless. This problem is caused by poor positioning and lack of precise 3D positioning information.

Due to the inadequate information support, device debugging, program preparation and debugging for machines spend a lot of time. This reduces machines' utilization rate. Although human create machines and most of machine equipments can achieve automatic control, manufacturing industry is still in the work age of human with limited information subordinating and serving machines.

The emergence of sensor technology, sensor networks, industrial wireless networks and new material, can bring human more necessary information to increase efficiency of human-machine interaction. The information for machine working targets can be obtained. The changing information for machine interior, state information of work environment, position and changing information for objects machined can also be obtained. A large number of key nodes' state for manufacturing systems and manufacturing environment are indicted with different forms. People not only process the information with screen, but also further request to be naturally integrated into system by vision, hearing, smell, touch as well as body, gestures, passwords, etc. With written interaction and voice interaction and so on, human-machine direct dialog can be realized. Informationized manufacturing based on ubiquitous perception in the age of human-machine interaction is coming.

Sensors can be installed on the machine tool bed and spindle to make people to know thermal field state of machine tool and deviation state of spindle. Therefore, the machine tool can be controlled automatically, the optimal working time and products quality can be ensured, and the efficiency of machine tool can be increased. The robot, its peripheral devices and its work space are installed accurate positioning sensors to ensure accurate positioning, and support automatic modification of programming, and reduce programming time. The rolling roll and motor in steel rolling are equipped by sensor to make operators to hold initiative in safe production and increasing efficiency. The information of machining environment, e.g. workshop temperature, humidity, electromagnetic sound and light, interacted state information among equipments interior, equipments and equipments, workpiece state and logistics information will be available for managers timely. Managers depend on view and language, etc. to make the decision under the condition of adequate and timely information. Manufacturing environment will enter an era of ubiquitous perception manufacturing, and propose new demands and new possibilities for human-machine interaction means.

Massive information is provided to equipment operators and production managers with various ways. At the same time, human will timely input efficient information into the automatic control system to improve the capability of accurate judgment and decision-making. However, adding implementation parts or even intelligent material to machines can make the equipments to adjust its behavior automatically. For example, automatic compensation for deviation caused by abrasion can urge machines more intelligent. Human-machine interaction is experiencing the transformation from precision to vagueness, from single-channel to multi-channel, and from 2D interaction to 3D interaction. The production equipments are going towards intelligent direction. The manufacturing activity will enter the era of human-machine compatibility work oriented human and hence enter higher stage of human-machine interaction, as shown in Figure 3.5.

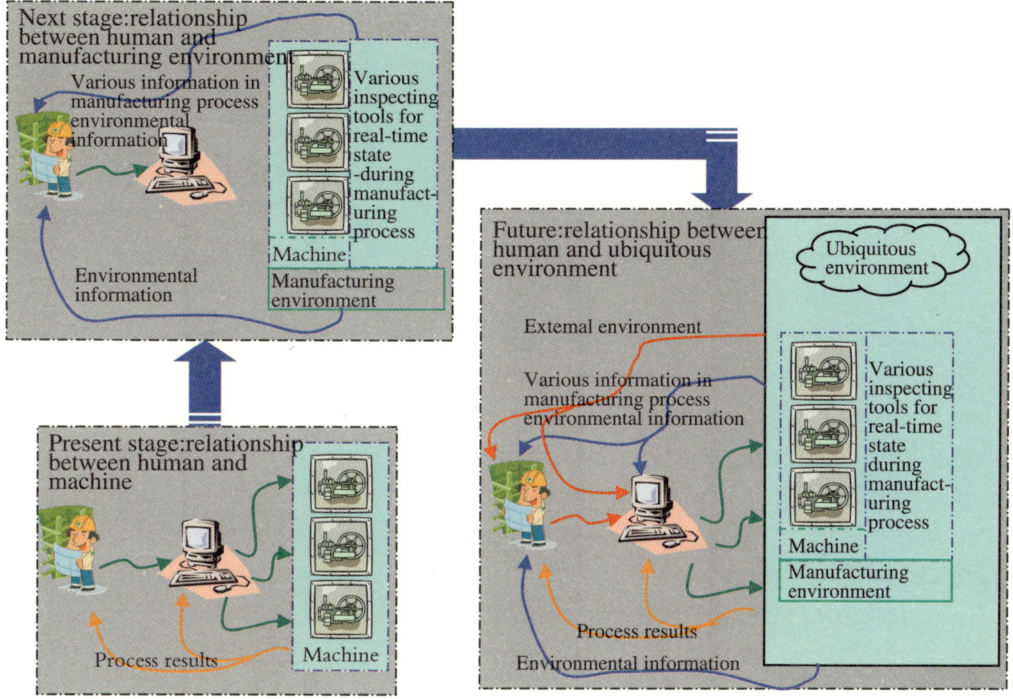


Figure 3.5   Development of human-machine interaction

## 3. Objectives and Tasks

The methods of human manipulating machine are experiencing the development from the stage of limited information to ubiquitous perception stage with rich information. The rich information further impels machine's control system to develop towards intelligent control. Human will control machines in a higher level. With the supports from related fields' technologies, manufacturing with limited information is going to informationized manufacturing based on ubiquitous perception, and then to various developing stages of intelligent manufacturing. A variety of human-machine interaction technologies will be emerging. The level of human-machine interaction will be promoted and human-machine interacted method will be improved.

## 4. Developing Roadmap of Human-machine Interacted Technology

At present, with inadequate information, human use operation and control system to control a loaded equipment or flow. With the development of information sources and communication technology, human will enter the stage of informationized manufacturing. Human will face many complex information sources, and face the whole production system to design control system and management approaches. With increasingly improved intelligence of production equipment, human will control production systems at a higher level.

Human-machine interaction in manufacturing system is to use the wisdom of human and operation and control means to control and manage production equipments and system to achieve production targets required. Under the premise of achieving targets, human continue researching more convenient and more efficient interacted means. In the beginning, human used hand to move switch, make punched tape and then use keyboard. The way of using hand was simplified gradually. However, the way of using hand is not the best mean for human, and thus mouth, eyes, ears will be used also. Obviously, the most convenient and most efficient way of interaction is: the machine would do what human want to do.

Programming will be the main means of human operating and controlling machine. In this period, the direct generation of machining codes by using CAD and CAM needs to be done. Based on the automatic programming for individual machining object, general and automatic generation of programming should be achieved by establishing increasingly rich factory database and embedding professional knowledge to provide technical information. With the supports of 3D high-precision positioning sensing system, industrial robots will achieve accurate positioning automatization, which enables robots to judge and adapt deviation (such as deviation of workpiece), to realize off-line programming, and to automatically adjust to generate new programming based on existing and similar programming. The robot programming efficiency will increase tens of times and reduce the professional requirements for user.

Multimedia feedback system, representing debugging process and production conditions, will be applied. Enhancing the capacity of the machine feedback is crucial for increasing work efficiency. During the process of debugging for equipments and procedures, view and sound, etc. can be used to reflect debugging information, for example animated simulation can reflect machining process of workpiece, and audio information can reflect machining state. Moreover, production process monitoring and fault prediction will also inform to human timely with language information.

Multimedia feedback system will endue managers with information feedback modes based on cognitive science. With the perception of environment on site, production process mastery, accurate report for machine interior, manager can decide easily and display their intelligence.

Virtual reality technology will not only be applied to the design, but also will become the primary means of program debugging. Workpieces to be machined and technical background are integrated with machining equipments on site to establish the link between workpieces to be machined and real machining scene. Debugging program in virtual machining will greatly reduce the occupancy time of the machine.

Robots are able to imitate human behavior. The sensors are installed on the operator's body to make the robots to perceive and know movements and posture of a part of human. Based on the context, the robots will understand and predict human behavior and understand human motivation to achieve humanoid action learning of robots.

With the breakthrough in speech recognition technologies, gestures, eye contact and language are needed to tell the robot how to work, as if the master tells the apprentice. The robots will become an assistant of human. With the convenient and effective interaction, human can solve the complex problems, while the robots will enlarge and extend people's ability and perform complex tasks and objectives.

With the development of brain science, human can control the machines and production equipments with consciousness. Now, human wearing special helmets have successfully made a call by using visual telephone dial in laboratory, as shown in Figure 3.6. With the progress of brain science technology, human's thought will be understood and become machines' language to control and instruct machines. This will be a revolutionary progress of human-machine interaction and will change the control method for machines, and even theory.

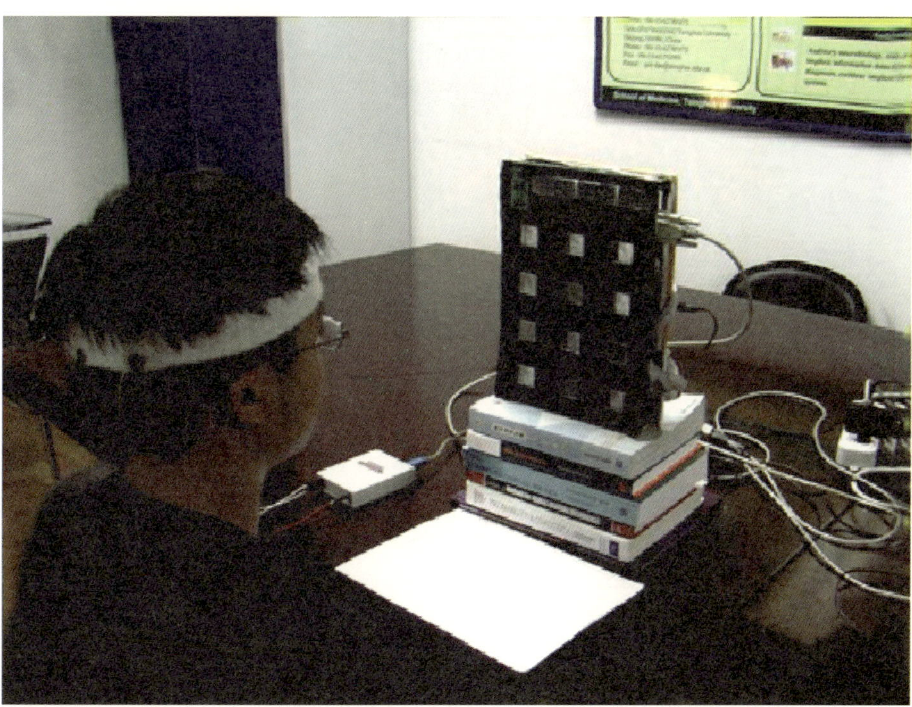

Figure 3.6  Brain consciousness realizing human-machine interaction  (source: http://news.tsinghua.edu.cn)

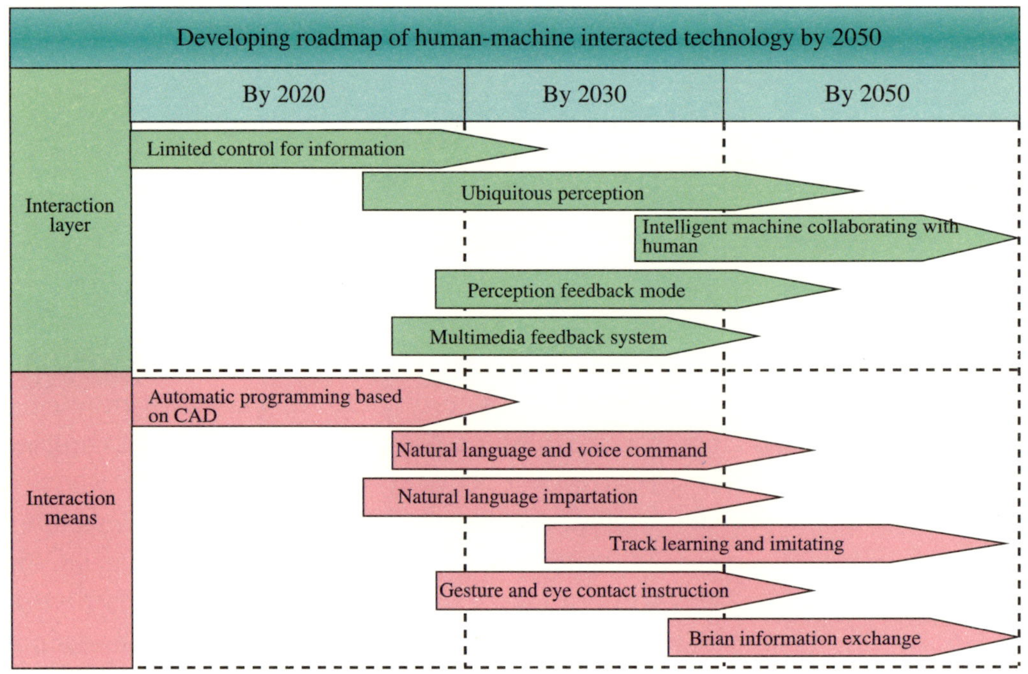



## 5. Summary

Due to the increasingly rich manufacturing information and advanced control means, manufacturing equipment is going towards intelligent direction. The level and means of human-machine interaction will develop in the direction of supporting human-machine compatibility.

# 3.5 Distributed Spatial Collaboration

## 1. Demands and Challenges

With the rapid development of technology and science in manufacturing engineering, many advanced technologies and new manufacturing modes are emerging to promote the market competitiveness of the manufactured products. A large number of intelligent manufacturing equipments are emerging, which take sensor and Radio Frequency Identification (RFID) as their cores. They become hot research topics of next generation manufacturing based on ubiquitous information. At the same time, current manufacturing industry is faced with huge challenges. For example, how to enable the seamless integration of manufacturing system and massive intelligent manufacturing equipments? how to provide highly collaborative effects to yield a new kind of manufacturing mode under ubiquitous information perception environment? how to respond to the changes of market demands rapidly and economically? and how to establish spatial collaboration manufacturing environment with virtual-real

   *Advanced Manufacturing Technology in China: A Roadmap to 2050*

integration? A manufacturing system should not only satisfy the requirements of low production cost and high product quality, it should also respond to the changing market's demands rapidly, adapt to the dynamic and changing physical manufacturing environment quickly, collaborate with massive intelligent equipments in physical manufacturing environment efficiently, and enable agile, adaptive and highly efficient manufacturing process. The development trend of manufacturing system technology is to integrate more and more network-based, distributed and intelligent equipments, in order to introduce production monitoring to the production cycle. The purposes are to reduce the installation, maintenance, operation time and costs of production equipments, and to improve the performance of manufacturing system. Therefore, the collaboration of network-based, distributed and intelligent equipments will play an increasingly important role in manufacturing system. A significant topic of future manufacturing industry is how to collaborate the network-based, distributed and intelligent equipments in special space so as to improve the agility, adaptability and efficiency of manufacturing system.

The future manufacturing environment will be a physical space embedded by the sensing, computing, control and information equipments. It will allow users to obtain information easily from physical spaces with the collaborative effects of network-based, distributed and intelligent equipments, and to make their correct decisions accordingly. Many key research topics arise in such circumstances. For example, how to establish collaborative model of various intelligent equipments? How to integrate equipments' information? how to process information semantically, intelligently and automatically? and consequently, how to control decision making in real time? On the above basis, in the circumstances of nondeterministic structure, an independent entity should be familiar with and adapt to its unknown environment quickly, and finish its work efficiently to achieve the desired targets. Under the coexisting condition of massive, heterogeneous and intelligent equipments, the key technologies of distributed spatial collaboration should breakthrough the limitations of current manufacturing system in information acquisition, monitoring, human-machine interaction and management, with the help of ubiquitous network. The agility, adaptability and efficiency of manufacturing system should be enhanced by the distributed collaboration of intelligent equipments.

Distributed (spatial) collaboration technology is a kind of integration technology oriented to specific manufacturing environment and specific application. In order to meet the requirements of advanced manufacturing, distributed collaboration technology are faced with the following great challenges:

(1) How to assure the supporting technologies of space collaboration meet the requirements of manufacturing systems?

(2) How to employ distributed spatial collaboration to achieve system monitoring, control, human-machine interaction as well as optimization of

production scheduling?

(3) How to use space collaboration in the integration of heterogeneous equipments, heterogeneous manufacturing modes and heterogeneous manufacturing systems to satisfy the requirements of agility, adaptability and efficiency of manufacturing system?

## 2. The State of the Art and Development Trends

Currently, the advanced manufacturing systems, both process manufacturing and discrete manufacturing, use the network-based system to manage and control production. In the layer of equipment and control system, production is monitored and controlled by the equipments' network and production process automation technology. In the layer of production execution, Manufacturing Execution System (MES) is used to finish production scheduling and optimization. In the layer of enterprise management, Enterprise Resource Planning (ERP) is utilized to achieve resource planning for enterprises. These technologies enable manufacturing industry to transform from the traditional massive production mode to new modes, like Lean Production (LP), Computer Integrated Manufacturing (CIM), Agile Manufacturing (AM), Global Manufacturing (GM), as well as the dispersed and network-based manufacturing.

However, due to the limitations of technological advancement, existing manufacturing systems have many deficiencies. In the layer of equipment and control system, there exist bottlenecks, as inaccurate, incomplete and inconsistent data can't be processed in real time. In the layer of production execution, most existing systems are using deterministic scheduling algorithm. As scheduling algorithm is hard to be optimized even under predictable conditions and it is difficult for current MES to collect the real-time production status, current MES can not easily adapt to the production changes and emergency events in the manufacturing workshops. In the layer of enterprise management, it is difficult for the existing information to be shared among present ERP systems and information systems in manufacturing workshops, and so the optimal allocation of enterprise resources is hampered. In the layer of production's operation, current human-machine interaction mode needs to be improved, especially under extreme production conditions.

In the environment with massive, networked, distributed and intelligent equipments, in order to satisfy the agility, adaptability, and efficiency requirements of production, distributed collaboration technology becomes the key point and trend of future manufacturing system, which is shown in the following aspects:

(1) The demands of industrial application drive the development of network-based, distributed and intelligent systems. At the moment, robots, made with mature embedded computers, can be used widely in such areas as vehicle manufacturing, equipment production lines, manufacturing operations & packaging and so on. In the future, robots will be integrated with more

    *Advanced Manufacturing Technology in China: A Roadmap to 2050*

intelligence, control, computing and communication capabilities. They will become a more and more important unit of manufacturing execution.

(2) Embedded systems integrated with sensing, data acquisition, wireless communication and control, will become the key of future process control and automation. The process control and automation equipments, which are made with the integrated and embedded systems, will replace some parts of the existing control units.

(3) Manufacturing processes of new products are becoming more and more complex, for example, production of silicon wafers needs a large number of sensors to acquire all kinds of data. Due to the flexibility of wireless sensor system, it will probably become the primary means of data acquisition in the future.

(4) Manufacturing systems need to achieve the integration of production monitoring, fault diagnosis, maintenance and control in the future. So they demand the breakthrough of the existing form of control data flow, which is "top-down". In fact, another kind of data flow form is also needed, which is from sensing to decision, or namely "down-top". These two kinds of data flows can work together to record the current production processes and equipment running status. Then, a new close-loop control can be established using intelligent sensor system and fault-tolerant control system.

(5) The developments of intelligent sensors, embedded computing, network technology, advanced control, intelligent technology and so on, enable manufacturing systems to acquire the required data and make correct decisions timely. Hence, the information gap among production monitoring, control, and management can be reduced.

## 3. Objectives and Tasks

The development objectives of distributed spatial collaboration technology are aiming to enhance agile, adaptive and efficient production of manufacturing systems, and to take ubiquitous network, human-machine interaction, ubiquitous information process and manufacturing system integration as its basis. Distributed collaboration technology will overcome the limitations of poor integration and weak collaboration in information acquisition, monitoring, control, human-machine interaction and management of current manufacturing systems. Meanwhile, the performances of manufacturing systems will be improved via the distributed collaborations among the massive and heterogeneous sensing equipments in manufacturing environments.

Based on the background of manufacturing systems, distributed collaboration technology should have new breakthroughs in its supporting technology, which is oriented to massive human-machine compatibility. The modeling and simulation technology of distributed spatial collaboration should have breakthroughs too. Research on information integration mechanism of distributed collaboration and scene analysis in manufacturing environments with human-machine compatibility needs to be conducted. This will lead to the formation of new collaborative manufacturing mode and computing model, with the span of spatial and temporal modes.

The collaborative control of network-based, distributed and intelligent equipments should be designed to achieve the integration of production monitoring, fault diagnosis, maintenance and control. The manufacturing process then, is supported by spatial cognition mechanism, with distributed sensing of production process status, real time acquisition of the sensed information, space semantic modeling and semantic collaborative computing. Further, production scheduling and optimization under the condition of large-scale ubiquitous sensing information acquisition, provides technical supports for parallel execution and the control and management of complex systems. Therefore, more effective and agile production process can be achieved.

The idea of collaborating in distributed spatial collaboration comes from multi-variable collaborative control theory and "decomposition-collaboration" method in large-scale system. Distributed collaboration helps us achieve the overall objectives and finish the tasks of multi-variable large-scale system through the collaborative running of multiple variables' control process in small systems. Therefore, distributed collaboration is also an important methodology for organization management and command scheduling. Distributed collaboration in those domains mainly includes human-machine collaboration, multi-model collaboration, multi-base cooperation, multimedia collaboration and so on.

### (1) Human-machine collaboration

In management field, human and machine (computer) have their own advantages and disadvantages respectively. For example, human's advantages include high intelligence, creativity, flexibility and initiative, while human's disadvantage is that human is easily influenced by subjective feelings and objective disturbances. In addition, long hours' work can lead to fatigue and weaken memory. Computer's advantage is that it can process massive information with high speed and precision. Moreover, it can also save and manage data effectively. Computer's disadvantages include its low intelligence level, poor creativity, initiative and flexibility. Hence, in the design of intelligent management system, the workloads of human and machine should be assigned reasonably so that the system will take the advantages of both, and human-machine intelligence should be integrated to achieve human-machine collaboration.

### (2) Multi-model collaboration

As an intelligent management system uses generalized models, a variety of models need to be collaborated, such as knowledge model, mathematical model and network model. Suitable models are used according to user's demands, and integrated models for solving practical problems are constructed in order to

   *Advanced Manufacturing Technology in China: A Roadmap to 2050*

adapt to the specific demands of various management activities.

### (3) Multi-library collaboration

An intelligent management system uses multi-library collaboration software, such as four libraries (i.e., database, knowledge, model and method) collaboration. In actual operation of multi-library software, the collaboration, scheduling and communication should be implemented. Each library can work independently and also run with multi-library collaboratively.

### (4) Multimedia collaboration

An intelligent management system uses human-machine interface with multimedia, which includes a variety of input devices, display screens and other output devices. Multimedia collaboration is needed to assure collaborative work of input devices and output devices, time sequence synchronization, and the scene matching of sound, graphics and textual information.

## 4. Development Roadmap

Distributed spatial collaboration work in virtual-real integrated manufacturing environment is an important developing direction of intelligent manufacturing with human-machine compatibility. It has important practical value for multi-perception equipments' collaboration and management, ubiquitous information integration, complex product design, simulation and training of complex systems and so on. In order to achieve the objectives of collaboration in manufacturing environments with massive, networked, distributed and intelligent equipments, current technology development should be combined with the development trend of manufacturing mode, so as to make great progress in the development of manufacturing prototype systems. Distributed spatial collaboration is an interdisciplinary research area in information science field. Thus, it should break through the limitations of traditional disciplines. Manufacturing systems in distributed collaboration environments need more mutual penetration of information science, manufacturing technology and management science.

### 1) Research on distributed collaboration theory and models for manufacturing applications.

The developments of sensor technology, micro electro-mechanical system, embedded computing, communication technology, control technology and intelligent technology bring the collaborative computing demands of trans-fields technologies into manufacturing environments. Constructing distributed collaboration models for manufacturing applications is the precondition and foundation of establishing an intelligent manufacturing system with human-machine compatibility.

(1) Modeling and optimization of distributed collaboration

Study heterogeneous equipments, heterogeneous manufacturing mode, heterogeneous manufacturing systems, as well as the corresponding communication modes, system monitoring and control, human-machine interaction and scheduling models which are used in distributed collaboration. The study can realize the mapping of artificial manufacturing system from physical space to the process of information networks. Scheduling optimization, multi-variable optimal control, computational intelligence, machine learning and so on, are used to optimize those models.

(2) Unit device for distributed collaboration

Develop interactive technologies based on ubiquitous perception and the method, technology and equipment of space positioning and gesture perception. Research on mobile equipments which are suitable for efficient and real-time drawings, the seamless integration technology and drawing tools in dynamic virtual environments to create high realism and multi-channel interactive technology and portable interactive equipment for virtual-real integration.

(3) Integration environments for distributed collaboration

Study system structure, interface, data processing and resource sharing in integration environments for distributed collaboration, mean while, design interfaces of application environments as well as practical and semi-practical equipments. Construct networks and toolkits for mobile interaction support, and integrate tools and equipments for project development. Form the open, service-oriented general environments.

**2) Distributed collaboration models for manufacturing systems**

(1) Distributed collaboration of ubiquitous networks and embedded systems

The embedded systems' requirements can be summarized as followings: different characteristics of QoS performance, including predictable delay, jitter and throughput; different service requirements of different components and environment conditions; coordination between service requirements of same and different levels; taking into account application requirements on initiative and real-time systems, robust adaptability for the requirements of dynamic systems and the conditions of dynamic environments. To support resource allocation, task implementation scheduling and high-level QoS requirements of distributed and real-time manufacturing systems, middleware is needed for the complex tasks. Under cable network condition, resource allocation, task implementation scheduling and high-level QoS requirements have been paid widely attention, but there has not been deep research under the condition of ubiquitous network so far. In order to satisfy the requirements of distributed and real-time manufacturing systems, the QoS demands should be met at system level of ubiquitous network, and abstract QoS model should be provided at application level.

(2) Network-physical system collaboration model to support manufacturing systems

Network-physical system (CPSs) is the integration of computing, communication and physical process. It surpasses those existing wireless sensor networks characterized mainly by perception, because it also emphasizes control and affect on the physical process after the physical world is perceived. Taking

     *Advanced Manufacturing Technology in China: A Roadmap to 2050*

into account the requirements of manufacturing systems and the characteristics of distributed collaboration, the distributed, real-time and embedded systems should consider the factors of physical process in manufacturing systems, and transfer to stable and reliable CPSs to meet the new requirements of manufacturing systems. The following four points should be taken into consideration.

①Concurrency of physical process and computing process in manufacturing systems.

②Distributed and network-based.

③Support the reconfiguration of manufacturing systems.

④The safe (hard) real-time responses of large-scale hybrid systems. This enables the CPSs' collaborative model of supporting manufacturing systems to bridge, connect and realize physical space and information space.

(3) Collaboration model for industrial wireless communication networks and human-machine interactive systems

In any manufacturing environment, a wireless-based manufacturing system has obviously advantages on data acquisition, sensing, monitoring, control and human-machine exchange and so on. It will provide technical support for achieving more agile manufacturing systems. The development of network-based, distributed, real-time and embedded systems and CPSs will further promote the development of large-scale wireless industrial networks. The emergence of large-scale, reliable and steady wireless industrial networks will bring tremendous changes for future management systems in terms of production process monitoring and control as well as production scheduling and management system. The particularity of manufacturing systems and the stability, reliability and security of wireless industrial networks should be considered primarily. Firstly, the robustness of interference and noise of wireless networks, as well as the robustness of manufacturing systems influenced by complex environments, should be improved. Secondly, stable connection in dynamic environments guarantees reliable networks and its QoS requirements under different conditions.

### 3) Breakthrough of bi-direction and trans-domain space collaboration modes

Data synchronization interface is developed among different levels of manufacturing processes and manufacturing spaces (abbreviated as "domain"). The interface can be used to automatically transfer in bi-direction the spatial collaboration data between upper and lower domains. The distributed and embedded data in equipment layer can perceive the information in control and production execution layer. Then, they will achieve abstract bi-direction data communication in the layer of enterprise management services. For the integration of equipment layer, control layer, production execution layer and enterprise management layer during manufacturing process, the impacts caused by network-based and distributed systems should be considered. The abstract data flow mode of spatial data collaboration from low-level data, through middle-level information, to high-level service, should also be taken into account.

(1) Integration of production monitoring, fault diagnosis, equipment maintenance and production control

Distributed, real-time and embedded systems are used to acquire and process production equipment data and production process data in real time. The running status of current production equipments and production processes can be recorded and reported. With the stability/reliability analysis of manufacturing systems under ubiquitous network condition and the agent-based network control technology, fault diagnosis and control of manufacturing systems can be achieved.

(2) Dynamic production execution scheduling for real-time feedback on workshop production changes and emergency events.

The running status of current production equipments and production processes is recorded and reported to reflect production changes and emergency events in manufacturing workshop. Based on the acquisition of real-time status data, production scheduling and optimization in production execution layer are carried out to improve the agility and flexibility of manufacturing systems.

### 4) Distributed collaboration in parallel management modes

With ubiquitous network and perception technology, distributed collaboration can reflect the status of production process timely and accurately, and instruct production flow agilely and effectively. It can also overcome the deficiency of existing manufacturing management systems, and provide technical support for parallel execution as well as the control and management of complex systems.

Manufacturing systems are a kind of complex systems including engineering factors and social factors at the same time. Parallel management systems can execute the comparison between actual manufacturing systems and artificial manufacturing systems obtained by the mapping from physical space (process) to information network process. Hence, it can complete control and management of actual manufacturing systems more effectively. Distributed spatial collaboration in parallel management mode is an effective control and management means for complex manufacturing systems. Therefore, distributed collaboration systems in parallel management mode need to be improved in such aspects as modeling, analyzing and parallel execution. The developing roadmap of various key technologies included in distributed spatial collaboration is shown as follows. According to technology development, the progressive relationships from bottom to top are also listed.

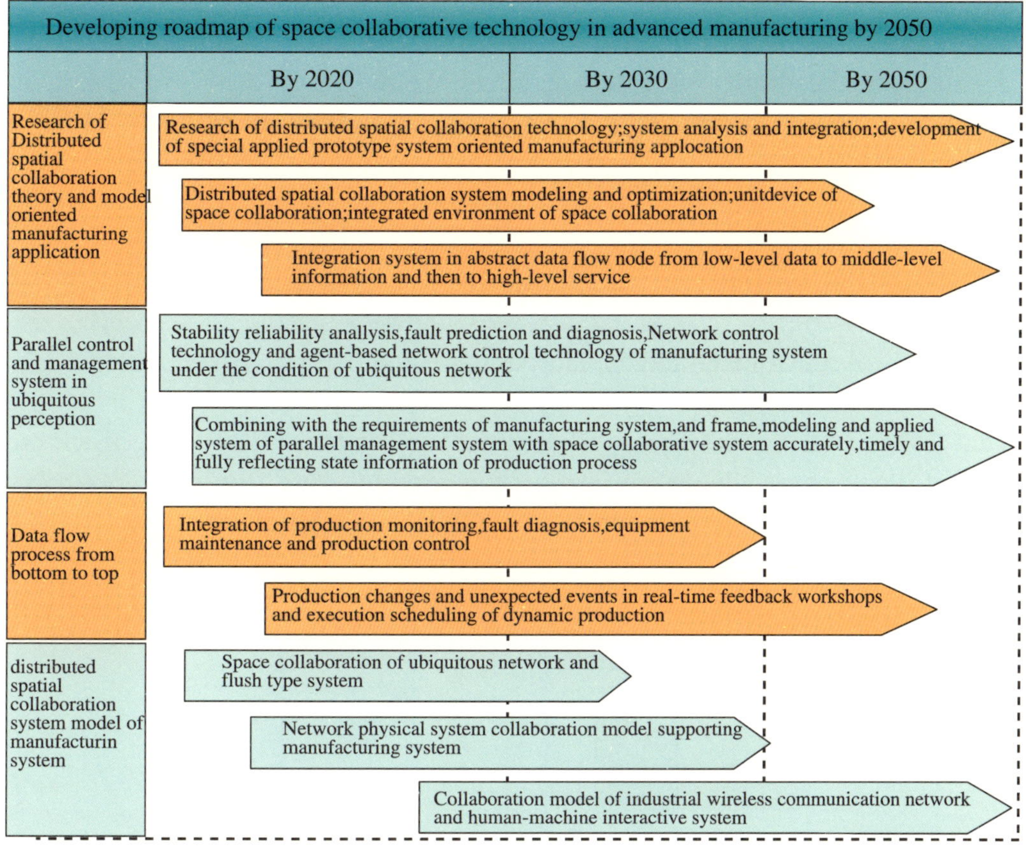


## 5. Summary

To meet the demands of market changes, manufacturing systems should not only pursue low production cost and high product quality, but also respond to the changing market's demands rapidly. In addition, it should adapt to the dynamic and changing manufacturing environments quickly. To sum up, networked, distributed and intelligent manufacturing systems will become the trend of future manufacturing industry. Moreover, the corresponding distributed spatial collaboration technology for manufacturing applications will enhance the agility, adaptability and efficiency of production in future manufacturing systems.

# 3.6  Parallel Management Technology

## 1.  Demands and Challenges

Parallel manufacturing systems consist of one actual manufacturing system and one or more corresponding artificial manufacturing system(s), as shown in Figure 3.7. Currently, advanced manufacturing industry takes digitalization, automation, networking and integration as its technical basis. Using parallel systems theory, parallel management technology refers to the related technologies that further promote the intelligent and scientific management level of advanced manufacturing industry. There are three main steps to build parallel manufacturing systems, namely A, C and P (abbreviated as ACP). The first step is A, that is, to construct an 'equivalent' artificial manufacturing system which consists of its social elements (humans, regulations and the environment, etc.) and its engineering elements (equipments, technique, materials and energy source, etc.). The second step is C, to recognize the interaction rules of various elements' changes using computational experiments method. On this basis, the last step P is to achieve parallel execution of one actual manufacturing system and its artificial manufacturing systems. There are different forms of parallel execution, such as learning and training, experiment and evaluation, management and control, decision-making and optimization and others. These forms are collectively called 'parallel management'.

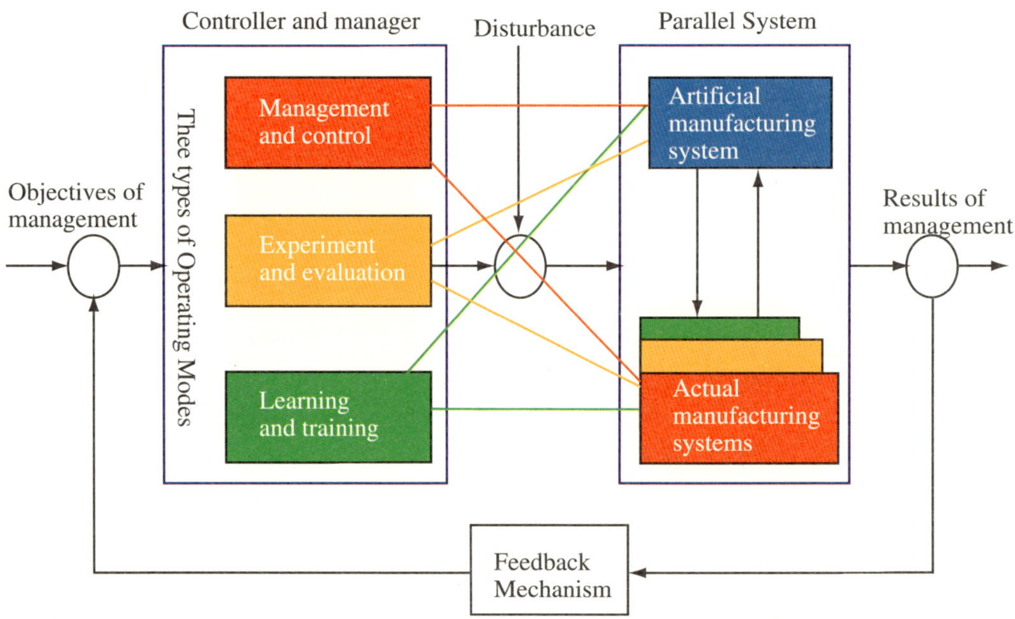


Figure 3.7   Frame of Parallel Manufacturing Systems

In the production layer of a manufacturing system, the influences of factors (e.g., personnel behaviors) on production safety, energy saving, environmental protection, efficiency, cost and so on, need to be considered. In the factory layer, the influences of economy, natural environment and customer demand changes on material supply, product sale and so on, need to be considered. In the enterprise layer, the impact of finance, market, policy, and politics, should be taken into account. The internal product design, manufacturing, marketing and services of an enterprise are more and more closely linked with its external industrial chain. Additionally, an abnormal state of any part of the whole industry chain can destroy the whole production system and result in severe consequences. Parallel management technology is such an efficient method that can possibly solve those problems. In normal circumstances, artificial manufacturing system uses computational experiments method to constantly promote the management level of actual manufacturing system, and yields better benefits. In abnormal circumstances, artificial manufacturing system assists actual manufacturing system to reduce and recover from its losses rapidly and correctly. Thus, parallel management technology continuously enhances the overall competitiveness of enterprises, and promotes a qualitative leap of the enterprise management level. It will become the core management technology for advanced manufacturing in future decades.

In order to satisfy the development demands of future manufacturing industry, parallel management technology is faced with enormous challenges. For instance, social and technical developments lead to more and more complex and fragile manufacturing systems, the increasingly rigorous demands on multiple objectives, and so on. Thus, parallel management needs to develop integration technologies of knowledge and intelligence, the expression, processing and application technology of human behavior, language, knowledge, wisdom and so on, and human-machine integration technology which organically integrate the collective wisdoms of group experts and artificial wisdom. Parallel management can still assure an advanced manufacturing enterprise to achieve its determined objectives, even though enterprises, customers, society and environments are changing all the time.

## 2. The State of the Art and Development Trends

At present, advanced process industry has basically implemented its IT system infrastructure with three layers for production automation, management informatization, and so on. In production equipment layer, Process Control System (PCS) is used for the optimal control and advanced control to achieve multiple objectives, such as high product quality and low production consumption. In production execution layer, Manufacturing Execution System (MES) is used to achieve the management informatization of production plan, quality, cost, energy source, materials, etc. In enterprise management layer, Enterprise Resource Planning (ERP) is used to achieve the management informatization of enterprise purchasing, production, sales, finance, personnel, and so on. They provide the technical basis for the standardization and modernization of enterprise management. However, enterprises are currently managed mainly with the experiences of management personnel, lacking of close loop control and management information systems which include management rules and humans in the loop. In the future, process industry will strengthen the integration of information systems and its business functions, add management informatization for its social elements, and create powerful, standard and unified information platforms. Information platforms will not only provide information and business functions, but also provide knowledge and decision-support functions. The intelligent management of enterprises will be further enhanced by means of human-machine integration.

Since 1990s, information technology has promoted the management informatization level of discrete manufacturing industry, in such phase as product design, manufacturing, marketing and service, and so on. With the formation and integration of global economy and global market, manufacturing industry gets its developing opportunities, and encounters great challenges meanwhile. In the past, much Computer Aided software (CAx) was developed separately for different purposes of discrete manufacturing industry. Currently, an integrated Product Data Management (PDM) is being developed mainly for product design. In the future, Product Lifecycle Management (PLM) will ultimately be developed for its full life cycle. In manufacturing industry, flexible manufacturing will be replaced by Design-For technology (DFx), for example, DFA (Design for Assembly). Network-based manufacturing on Internet platform will gradually become the dominant manufacturing mode.

In summary, horizontally all the manufacturing sectors expands their internal integration to the external integration with their suppliers, customers and other industrial chain. Informatization enterprise will organically integrate with e-commerce, e-government and informatization society. Vertically, the information integration and functional integration of manufacturing enterprises will extend to knowledge integration and wisdom integration. More and more attention will be taken to manufacturing compatibility with human being and environment, high production efficiency and speed, energy saving and low material consumption, and green production for environmental protection .

## 3. Objectives and Tasks

As an emerging technology, the development of parallel management technology will be oriented by the demands of manufacturing industry in future decades, and its driven force is to solve more and more relevant scientific and technical problems. Manufacturing science, management science, complex science and intelligence science are constantly integrated to obtain significant achievements in the areas such as parallel management theory, technology and system development, and so on. Parallel management technology will be applied to different phases and different levels of advanced manufacturing, to promote the enterprise's capability, discover complex changing laws, to recognize and prevent abnormal states, and to continuously increase the intelligent and scientific management level of advanced manufacturing enterprises. In order to achieve these objectives, R&D of parallel management includes the following primary tasks.

### 1) Artificial system technology

(1) Artificial manufacturing system modeling technology with human-
machine integration

Human-machine integration takes the advantages of nature humans (groups) and artificial systems to build up complementary systems. From easy to difficult, the system includes various modeling technologies for behaviors, languages, knowledge and wisdom, and so on. Behavior modeling mainly studies cellular automata model and other agent models. Language modeling mainly studies the language dynamics expression and word calculation method. Petri network model, translator model and its deriving models are used for agent interactions. Intelligent learning algorithms and game theory are used for agent-based decision-making. Simulation technology is used for the whole manufacturing process. Knowledge modeling technology mainly studies the expressing, storing, analyzing and applying of knowledge for human-machine integration. Wisdom is in the higher level than knowledge, and consequently wisdom modeling technology mainly studies the expressing, storing, analyzing and applying of wisdom for human-machine integration.

(2) General construction and validation methods of artificial manufacturing systems

The main tasks includes human modeling, social organization modeling, domain strategy modeling, social criteria modeling and spatial-temporal evolutionary modeling of artificial manufacturing systems; general construction methods and procedures of artificial manufacturing systems; the completeness degree, availability degree and 'equivalence' validating methods for artificial manufacturing system model.

### 2) Computational experiments technology

The main tasks includes the general strategy formulation and analysis of computational experiments executed in artificial manufacturing systems; the strategy design, calibration, sensitivity analysis methods and validation algorithms for computational experiments; two kinds of evaluation methods for computational experiments in artificial manufacturing systems (i.e., the evaluation method driven by objectives and the evaluation method driven by events); evaluation method with qualitative fuzzy experiments; evolutionary game approach; the results analysis method of computational experiments to find system elements' interactions and evolutionary rules.

### 3) Parallel execution technology

It mainly includes the general methods of parallel execution in complex manufacturing systems, such as interaction protocol and the 'equivalence' validation method of artificial manufacturing system and actual manufacturing system; controllability, stability definition of parallel systems and their judgment criteria; group strategy and optimization strategy of parallel systems; the evaluation criteria setup of multiple objectives and multiple effective solutions for parallel systems; the internal feedback mechanism for parallel systems and the corresponding adaptive control algorithm; novel optimization method of parallel systems, which is based on perturbation analysis and sequential optimization.

**4) Parallel system technology**

It mainly includes technical architecture and functional architecture, for example: Service-Oriented Architecture (SOA), Software as a Service (SAAS), and so on; technologies, standards and products relevant to Internet, information integration and integrated platforms; integrated modeling and overall optimization tools for complex manufacturing systems; computational experiment technology and parallel management technology for human-machine integration; system development languages, development tools, configuration tools, test and maintenance tools, infrastructure components, middleware, visualization tools, and so on. Parallel computing technology includes general parallel computing algorithms, special parallel computing algorithms which are based on CPU (Central Processing Unit) plus GPU (Graphic Processing Unit), etc.

**5) Parallel system platforms**

Parallel system platforms include system platforms with integrity, standards, good scalability and configuration. The construction of parallel system platforms will need to design an independent data processing module, support the identification and application of Internet resources, and organically integrate Internet resources to meet the requirements of software platforms. Moreover, it will design and develop databases, knowledge bases, basic component bases, artificial system constructing tools, computational experiments modules, parallel management modules, decision-making support tools and human-machine interaction interfaces, and so on.

## 4. Development Roadmap

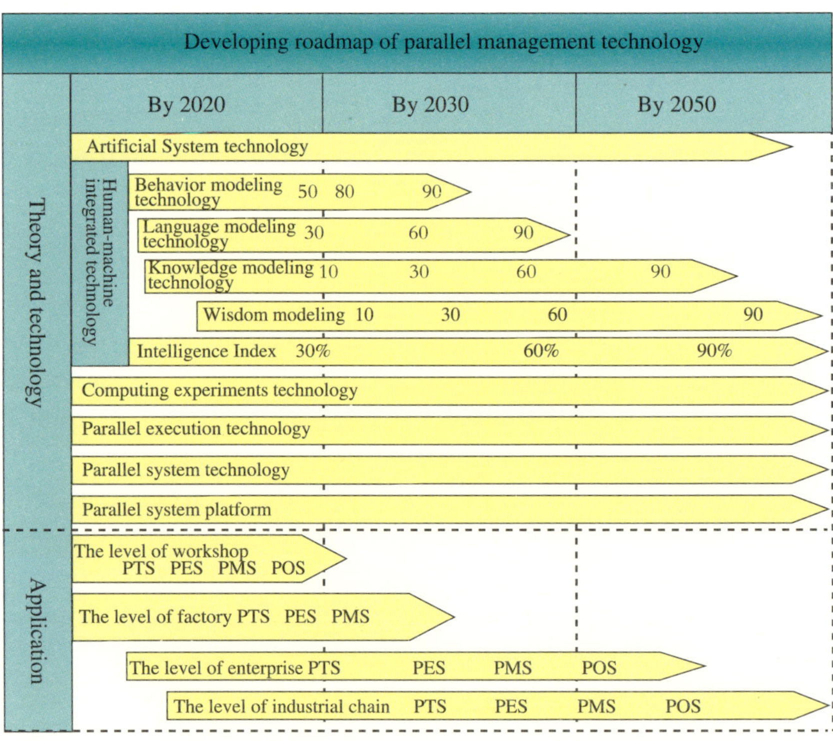

The developing roadmap of parallel management technology is designed according to the prediction of the manufacturing industry's demands in various periods in the future and the maturity degrees of their relevant supporting technologies.

**(1) Application Development of Parallel Management**

In the application layer of advanced manufacturing, parallel management systems will be developed for the workshop layer, factory layer, enterprise layer and finally industrial chain layer, according to the sequence from easy to difficult. To realize different application objectives of advanced manufacturing, according to the sequence from easy to difficult, PTS (parallel training system) will be developed for the purpose of learning and training; PES (parallel evaluation system) will be developed for the purpose of testing and evaluation; PMS (parallel management system) will be developed for the purpose of management and control; POS (parallel optimization system) will be developed for the purpose of decision-making and optimization respectively. According to the time sequence of its application in different industries, parallel management will be applied to process industries such as ethylene, electricity, coke, petrochemical, pharmaceutical, metallurgy, paper making, etc., and discrete industries such as machinery, equipment, automobiles, electronics, household appliances and so on. For application scalability, currently parallel management starts from individual project's development. Additionally, it will extend to mature product development for the same industry, and finally build up the software industry of parallel management which can be easily applied to different industries, and promote or even replace those PCS, MES, ERP, SCM (Supply Chain Management), CRM (Customer Relationship Management) systems being applied in enterprises informatization.

**(2) Technology development of parallel management**

There are five main kinds of parallel management technologies, which include artificial systems, computational experiments, parallel execution, parallel technology and parallel system. There are four typical application layers of parallel management, which include workshop, factory, enterprise and industrial chain. And there are four kinds of typical applications of parallel management, which include PTS, PES, PMS and POS (Table 3.1). The scalability and precision estimation, for the five kinds of technologies needed by the four kinds of applications in the four kinds of layers, are shown in Table 3.1. The scalability and precision value needed by enterprise POS is calibrated as its standard value of 100. For the precision and maturity degree of parallel system technology, the scalability and precision value needed by enterprise POS is calibrated as its standard value of 100.

| | PTS | PES | PMS | POS |
|---|---|---|---|---|
| Workshop layer | 10 | 20 | 35 | 40 |
| Factory layer | 20 | 35 | 50 | 60 |
| Enterprise layer | 30 | 50 | 65 | 80 |
| Industrial chain layer | 40 | 60 | 80 | 100 |

### (3) The intelligence-oriented development of advanced manufacturing using parallel management

To 2020, 60% of the data and information in advanced manufacturing enterprises will be processed by parallel management systems, while 40% of them will still be performed by human. 30% of the knowledge processing and intelligent decision-making tasks will be implemented by parallel management systems, while 70% of them will still be completed by human.

To 2030, 95% of the data and information in advanced manufacturing enterprise will be processed by parallel management systems, while 5% of them will still be performed by human. 60% of the knowledge processing and intelligent decision-making tasks will be implemented by parallel management systems, while 40% of them will still be completed by human.

To 2030, 99% of the data and information in advanced manufacturing enterprise will be processed by parallel management systems, while 1% of them will still be performed by human. 90% of the knowledge processing and intelligent decision-making tasks will be implemented by parallel management systems, while 10% of them will still be completed by human.

## 5. Summary

With the continuous integration and sublimation process of group wisdom and artificial wisdom, parallel management technology will help to recognize more and more complex changing laws of advanced manufacturing. Thus, it can deal with optimization challenge under ever-changing environments and emergence management challenge under abnormal conditions. Moreover, it will enhance its management level with the help of intelligent orientation and scientific orientation. If the developing roadmap is implemented accordingly, parallel management technology can be applied to different layers and different application areas of various manufacturing industries in order to meet the demands of enterprise management in different periods.

# 3.7  E-Business

## 1.  Demands and Challenges

In the traditional manufacturing industry, manufacturing procedure and marketing procedure are completely separated. However, with the revolutionary development of information technology and Internet technology, the integration trend of manufacturing procedure and marketing procedure has emerged. In this circumstance, modern manufacturing industry and its business mode are therefore facing the unprecedented challenges in the following aspects. First, in terms of its organization and structure, the socialization trend of production process leads to labor division and cooperation between manufacturing sectors, and industrial structure needs to be simplified and refined urgently. The 'pyramid' type of current enterprise structure gradually exposes its deficiencies, such as organization bloating, inefficient operations and so on. Second, in terms of its production process, manufactured product and its production process are becoming more and more complex. Its specialization degree and its production cost are increasing continuously. Third, in terms of manufacturing environment, fixed computing and production equipments are gradually fading out from production environments, while ubiquitous computing environments are emerging, which are based on mobile and ubiquitous equipments. Fourth, in terms of customer relations, Web 2.0 technology enables the diversification of customers' demands, and their participation interest is increasing gradually. Finally, in terms of market decision-making, the trend of economy globalization has resulted in more dynamic and complex market competition.

E-business, which emerges and is rapidly developed in 21st century, provides one solution to the reform and development of modern manufacturing industry. Generally speaking, E-business uses computers, networks and communication technologies to promote the electrification of manufacturing industry and its commercial activities, and involves such function modules as Business Process Modeling (BPM), Enterprise Resource Planning (ERP), online marketing, Supply Chain Management (SCM), Customer Relationship Management (CRM), Product Lifecycle Management (PLM) and so on. Among them, SCM, CRM and PLM compose the value-added chain of advanced manufacturing industry, and they help promote direct and transparent business procedure, and in turn increase manufacturing efficiency and business opportunities, so as to achieve the innovation of manufacturing and business mode. The advanced manufacturing industry, which is integrated with E-business, will help the formation of network-based 'virtual manufacturing enterprise', and contributes to resource sharing and cost saving. This not only increases the core competitiveness of an enterprise, but also enhances its self-adaptability for its complex and dynamic competition environments.

To sum up, the network-based, automatic and intelligence-based manufacturing industry integrated with e-business is the developing trend of advanced manufacturing technology in the next 50 years.

## 2. The State of the Art and Development Trends

The development history of manufacturing industry has mainly experienced three stages, namely traditional manufacturing industry before

1880s, computer-aided manufacturing industry characterized by such techniques as Computer Aided Design / Computer Aided Manufacturing (CAD/CAM) and Computer Supported Cooperative Work (CSCW) from 1880s to the late 20th century, and modern manufacturing industry started from new millennium. Modern manufacturing industry has initially integrated with E-business technologies, such as ERP, CRM and so on. So far, production procedure and marketing procedure in many modern manufacturing enterprises are still in the disconnection state.

With more and more manufacturing enterprises positively developing e-business technologies, the integration of manufacturing procedure and business procedure has become more and more evident. Two trends of integration, vertical integration and horizontal integration, have been shown.

### 1) Vertical integration

E-business organically integrates various key stages and procedures in product life cycle, such as new product creation, design, manufacturing, marketing and CRM. E-business uses supply chain to closely link factory-level enterprises for forming one manufacturing industry chain. It also enables intercommunication and co-development between the upstream and downstream enterprises in the supply chain, so as to realize the transformation from supply chain, through value chain, and finally to ecological chain.

### 2) Horizontal integration

E-business enables marketing procedure to break through the existing spatial-temporal limitation, and meanwhile makes manufacturing enterprises to face global opportunities and competitions more directly. E-business also changes the relationships among enterprise organizations. Currently, supply, production, distribution and retail enterprises are remotely distributed in geography, and are equal and independent in organization. E-business helps them establish cooperative relationships closely based on equal negotiations, and form one dynamic "virtual enterprise" or "enterprise alliance". The new organization form can achieve optimization, dynamic arrangement and sharing of enterprises' resources, and so represent the development direction of advanced manufacturing industry.

## 3. Objectives and Tasks

As far as the strategic development demands of future manufacturing industry are determined, government, enterprises and research institutions should focus on exploring and initiating the emerging E-business modes and its key technologies to enhance the competiveness of manufacturing industry. Considering the current states and future challenges that modern manufacturing industry faces, the overall development objectives of future manufacturing industry include establishing and improving the basis framework of key technologies for E-business and computational economics of advanced manufacturing industry; positively developing computational business which can promote the decision-making capability and adaptability of manufacturing enterprises when facing diverse and initiative users as well as dynamic markets; studying the new lifecycle management mode of products to adapt to the future ubiquitous computing environments; and constructing advanced manufacturing industry which is harmonious and sustainably developing. Considering the demands and challenges that modern manufacturing industry and its business mode are facing, the development tasks of future manufacturing industry integrated with e-business include the following.

(1) In terms of organization structure, it is needed to achieve network-based organization and collaborative manufacturing. The former is an organization in wich enterprises are link e-business on Internet so as to realize collaborative manufacturing and marketing. The latter is to achieve inter-enterprise design and manufacturing resource sharing using supply chain management. Both of them will greatly enhance the core competitiveness of an enterprise, and promote its advantages in group and industry layers.

(2) In terms of production process, it is needed to achieve Business Process Model (BPM) integration of advanced manufacturing industry. By using E-business, those heterogeneous subsystems, like SCM, PLM, ERP and CRM, can be integrated and the key stages' boundaries of product life cycle can be overcome. Thus, supply chain can become more smooth and efficient, and guarantee fast and real-time response to market changes.

(3) In terms of manufacturing environment, new ubiquitous equipments and network technologies should be positively developed, to achieve ubiquitous-oriented manufacturing equipments and production environments, as well as to establish the new lifecycle management mode of product in ubiquitous environments. This will standardize the ubiquitous manufacturing environments, procedures and its business modes, and thus bring about a new manufacturing revolution.

(4) In terms of customer relations, collaborative development and massive customization marketing should be achieved. The former abandons the old mode of 'work behind closed doors', and achieves product optimization and innovation using interactive marketing between enterprises and customers. The latter treats each customer as a potential sub-division market, and implements the customized product design, manufacturing and delivery according to the customized demands. Both of them are the preconditions and foundations to achieve customization marketing.

(5) In terms of market decision-making, it is needed to build up computational business and marketing model for advanced manufacturing industry. The computational business fully integrates existing research results from artificial intelligence, computational economics, behavioral psychology and E-business through the integration and innovation of concepts and technologies, in order to solve the problems of real-time optimization and decision-making in complex and dynamic market environments, and to establish the foundation for manufacturing industry as well as to deal with its

global competitions flexibly.

## 4. Development Roadmap

According to the strategic needs of future advanced manufacturing, the development roadmap of E-business can be divided into three different periods, namely 2010~2020 (short-term), 2021~2030 (medium-term), 2031~2050 (long-term).

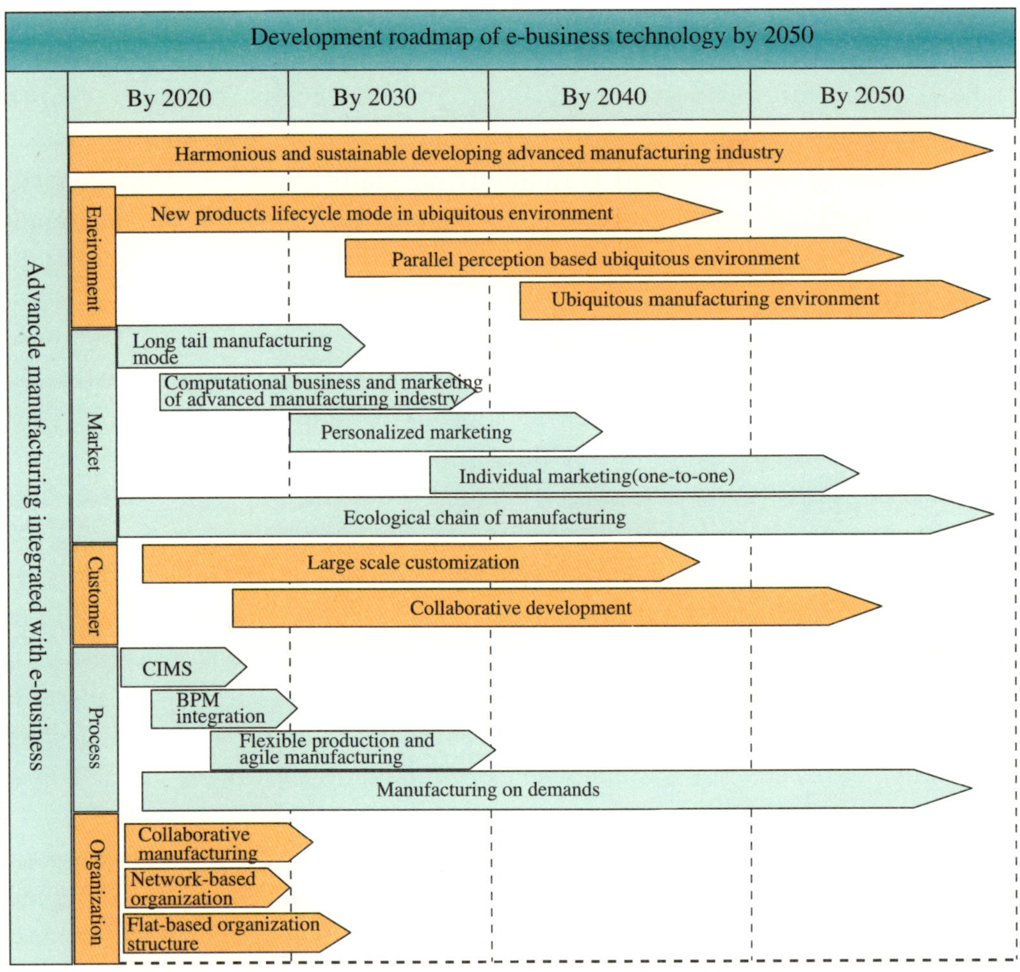


From 2010 to 2020, this period will be characterized by the informatization and networked virtualization of manufacturing industry. Manufacturing industry will achieve the reform and innovation of its organization structure and production process. The details are as follows.

(1) Traditional organization structure depends on hierarchical control. However, it needs to be adjusted properly at the moment. The flat organization structure in manufacturing enterprises should be applied gradually.

(2) The trend of global economy integration should be conformed, so the network-based organization with characters of intensive multilateral connection and cooperation should be established. Stable manufacturing enterprise alliance or dynamic virtual manufacturing enterprises should be formed.

(3) Based on E-business and computational economics platforms, those

   *Advanced Manufacturing Technology in China: A Roadmap to 2050*

upstream and downstream enterprises will achieve 'collaborative manufacturing' mode which is linked by the same supply chain.

(4) The Contemporary Integrated Manufacturing System (CIMS) integrated with E-business, and its key technologies should be actively developed. With the integrated optimization of different phases, such as information integration, process integration and inter-enterprise integration, the key aspects of manufacturing industry BPM, such as originality, design, manufacturing, quality assurance, marketing, sales and CRM and so on, should be integrated at a deeper level.

From 2021 to 2030, this period will be characterized by the personalization and informatization of the marketing mode of manufacturing industry. Manufacturing industry will transfer from traditional 'product-centric' marketing mode to 'customer-centric' intelligent marketing mode. The details are as follows.

(1) Based on the development of e-business and Web 2.0 technology, customers are guided to participate in the key aspects of BPM. According to their interactions and feedbacks, the improvement and innovation of new product can be completed, and the 'collaborative development' of manufacturing enterprises and customers can be realized.

(2) According to the sub-division needs of markets, marketing mode transformation from mass production to customization production should be achieved. The key technologies, like flexible production and agile manufacturing, should be researched and solved to realize the marketing mode of large-scale customization.

(3) Based on the integration of manufacturing industry BPM, computational business mode of advanced manufacturing industry should be achieved to realize dynamic market prediction, computational marketing and business game analysis and so on.

(4) Long tail manufacturing mode should be explored to enlarge the niche market of manufacturing industry, to reduce the barriers of manufacturing industry, and to promote the flexibility of manufacturing enterprises.

From 2031 to 2050, this period will be characterized by ubiquitous-based product manufacturing, management and business mode of manufacturing industry. Ubiquitous computing and mobile business technology will be widely used in manufacturing industry, so those manufacturing and business platforms based on new computing equipments will be emerging. Accordingly, manufacturing industry will establish its new lifecycle management mode of product to adapt to future ubiquitous computing and business environments. The details are as follows.

(1) Product identification, production line monitoring and storage management in ubiquitous manufacturing environments will be realized. The business procedures will be simplified and integrated.

(2) Important product parameters in supply chain and logistics system of ubiquitous manufacturing and business environments should be monitored in

real-time, and analyzed to predict and warn product status in real time.

(3) Parallel cognition in ubiquitous environment will instruct the marketing mechanism and procedure of manufacturing industry and accurately recommend relevant products to the dedicated customers, in order to truly achieve personalized marketing in manufacturing industry market.

(4) New lifecycle management model of product in ubiquitous manufacturing should be developed to standardize future manufacturing environments, procedures and modes.

To 2050, it is predicted that large scale manufacturing and marketing will be replaced by manufacturing and marketing adaptable to dynamic manufacturing demands, infinite sub-divisional markets, increasing diversity of customers' demands, ubiquitous manufacturing and business environment, integration of manufacturing and business procedure, seamless collaboration trend of different organizations, on-demand manufacturing and individual marketing mode . The details are as follows.

(1) Production and business philosophy in manufacturing industry will transfer from traditional 'product-driven consumption' to new 'demand-driven products'. Both manufacturing BPM processes and supply chain operations will be driven by individual customer's needs (there is no group customer's needs any more).

(2) On-demand manufacturing and business mode enables manufacturing industry to realize zero inventory management and optimize resource allocation.

(3) The close integration trend of manufacturing and business procedure will continue.

(4) Co-evolutionary ecological chain among manufacturing enterprises will replace current supply chain, and the ecological environment of mature manufacturing should be formed.

## 5. Summary

The advanced manufacturing and marketing mode, which is integrated with E-business, is the unique way to achieve the modernization of future manufacturing industry. It not only establishes solid foundation to achieve network-based organization structure, production process automation, personalization of customer relationship, intelligent decision-making of marketing and ubiquitous manufacturing environment, but also yields its positive and profound influences on constructing the harmonious and sustainable development of the ecological environment of advanced manufacturing, and on promoting the sustainable development and prosperity of the national economy.

# 3.8  System Integrated Manufacturing

## 1.  Demands and Challenges

System Integrated manufacturing is a kind of human-machine integrated system composed by intelligent machines and human experts. It is a kind of industrial manufacturing system of human-machine coexistence and collaboration. Under the support of computer network, business activities through whole life cycle of products, such as product design, material purchase, manufacturing, sale and use of products, breaks through the constraints of space and place for enterprise. Through the simulation of human's intelligence activities, the operating states of manufacturing system are monitored, adjusted and optimized, many unit systems, enterprise functions and processes are integrated to become an organic object with self-organization, self-learning, self-tuning and adaptive capacity. Thus, a part or whole of mental labor in manufacturing system is replaced, and intelligent management and control in all aspects of production process can be achieved, Intelligent management system (IMS) is a new generation of computer management system. Based on the functional and technical integration of management information system, office automation system, decision support system, it uses modern scientific methods and technologies, such as artificial intelligent expert system, knowledge engineering, pattern recognition and artificial neural network, to achieve integration, collaboration and intelligentization. Therefore, intelligent management system is also called intelligent integrated management system (IIMS) or intelligent integrated coordinating management system (IICMS).

In each aspect of manufacturing enterprises, such as operation, management, product development and manufacturing, information technology, modern design and management technology have changed the design method, design process and design philosophy of products, and promoted the innovation of manufacturing technologies and management models, as well as the innovation of inter-enterprise collaborative relationship. Product quality, productivity and innovation capability of enterprises can also be enhanced, and product design and production cycle can be significantly shortened. Energy consumption and material consumption can be declined and comprehensive competitiveness of enterprise can be significantly improved.

Due to the pressure of faster product update, continuous changing market needs, increasing impacts caused by global manufacturing, enterprises should focus on their core business to achieve advantages complementary, optimization, dynamic combination and sharing of resource. Integrated network can provide cross-region operating environment for design, production, management and marketing in manufacturing enterprises, realize global, integrated and ordered manufacturing industry. For product design, manufacturing and production management and other activities, as well as enterprise's entire business processes, the relating manufacturing resources should be fully and rapidly mobilized, organically integrated and efficiently used. Intelligentization will greatly improve the ability of adaptation to environment, the ability of processing massive and incomplete information, and the ability of coordination and collaboration for manufacturing enterprise, system and unit (equipment), and enable enterprises to adapt to changeable market demands and complex environments with fierce competition.

## 2. The State of the Art and Development Trends

System integrated manufacturing mainly involve intelligent technology, network-based manufacturing technology and system integrated technology. Its development process and phases are shown in Figure 3.8.

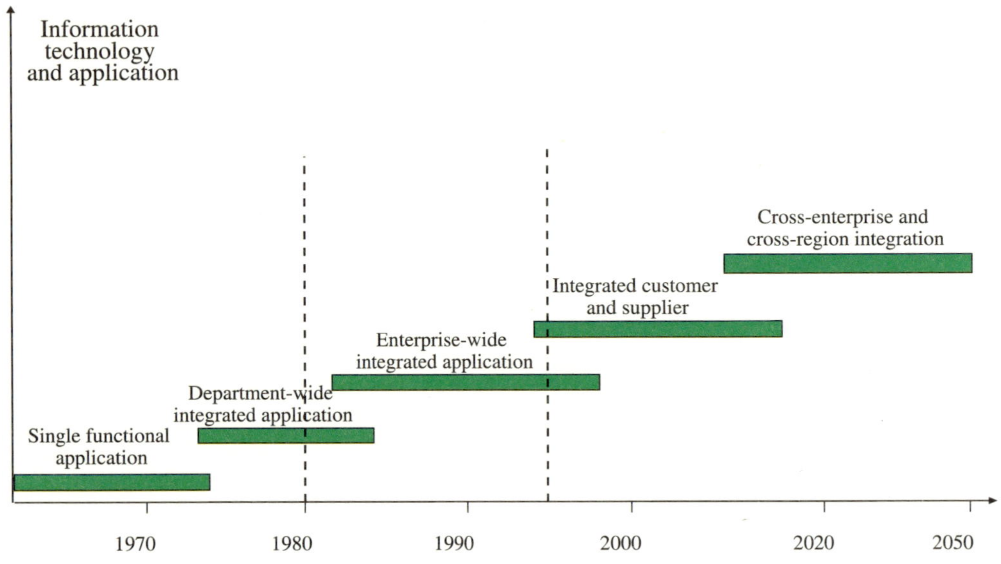


Figure 3.8  Developing phases of manufacturing system

Modeling method, fuzzy logic control, neural network control, expert system, learning control, hierarchical control, genetic algorithm of complex system are used to solve control problem of complex system with uncertainty model, high degree of non-linear, complex tasks requirements, etc. This can realize identification, memory and learning for information provided by an unknown system. And the accumulated experiences are used to improve their performance and the performance of the changed system is better than original system's. For complex tasks and distributed sensor information, self-organization and self-coordination functions enable system to be initiative and flexible, and to have human-imitating intelligent operation mode and multi-objective optimization process. However, the application range is still being widened, timely control capability is being improved, and the bottleneck of knowledge acquisition and optimization are being solved at present. The integration of various intelligent control methods is mainly limited to individual

   *Advanced Manufacturing Technology in China: A Roadmap to 2050*

control system.

With the rapid development of computer and network technology, manufacturing system has undergone the following phases: the single machine application phase which can complete independent, local computing and management function，the unit application phase which can complete associated business within and between enterprise's departments，enterprise integrated application phase which can complete integration and optimization operation of enterprise's all unit application, and achieve integrated management of product design, process design, manufacturing resource management, manufacturing process management, product sales, after service and other business, and the inter-enterprise integrated application phase supported by wide area network and Internet, and the enterprise informationization still has a long way to go.

Artificial Intelligence merges into all aspects of manufacturing process, and enables system to automatically monitor its running state. When system is stimulated by external or internal motivation，it can automatically adjust its parameter to reach optimal working condition and have self-organizing capacity. New rapid identification, optimization, and self-learning control method, real-time intelligentization and knowledge intensive artificial intelligence by means of knowledge process are playing increasingly important role. Fuzzy logic neural network composed by self-learning of neural networks and fuzzy logic reasoning with better knowledge express ability is still an important issue of the development of intelligent control. The high degree integration of mechanical science, life science, information science, materials science enables bionic manufacturing, which can imitate functional structure and operation mode of biological organs' self-organization, self-healing, self-growth and self-evolution and so on, to extend the organization structure and evolutionary process of human. Up to now, advanced manufacturing management system can not adapt to objective needs of development because of low intelligent level. The main problems are listed as follows:

### 1）Low intelligence of problem description

The descriptions of actual problems rely merely on data models, while for some non-formulaic, non-structured or semi-structured problems, data models can not describe or is difficult to describe them. For example, for the description of management decision-making which needs experts' experience and knowledge, data model is hardly competent.

### 2）Low intelligence of problem solving

The singleness of description method for problems must limit the way of problem solving. When problems are being solved, mathematical calculated method based on mathematical model can only be used to solve those mathematical models which can be solved in math. However for those mathematical models which can not be solved in math, there is no way to solve. For example, some problems dependent on knowledge description and logic judge such as fault diagnosis and fault handling of production equipment.

### 3 ) Low intelligence of information process

For information processing，only database technology is adopted, which is mainly used to process numeric information, while it is difficult to process massive knowledge information, inquiry and call of various models and methods needed by management activities process. This directly influences quality and efficiency of management.

### 4 ) Low intelligence of human-machine interaction

Human-machine interface is non-intuitive, unfriendly and inconvenient, and information input and output is difficult, so non-computer professionals are difficult to use it . The division of human and machine is unreasonable, efficient computer aided management can not be achieved. The optimal integration of human and machine intelligence is absent, and human-machine disorders.

### 5 ) Low intelligence of system maintenance

Due to the deficiency of efficient software maintenance, updates, expansion capacity, as well as self-organizing, adaptive, self-learning function, development cycle of system is longer and can not play its roles, and results in shorter system lifecycle.

With the support of continuous development of modern manufacturing mode, the integration of many kinds of technologies, e.g. information technology, management technology, design and technics technology, enables manufacturing system to break the boundaries of enterprises. Product design, material selection, parts manufacturing, market development and product sales can be carried out in different places or different countries. Multi-disciplinary and multi-functional integrated product design can achieve optimal combination of products dynamic and static characteristics, efficiency, precision, life, reliability, manufacturing costs and cycle. The integration of CAD/CAM/CNC can  complete modeling, programming, modify, and CNC machining on site for complex components and achieve multi-task parallel processing, then rapid and efficient manufacturing of materials and parts, as well as their integration can be obtained. Considering the optimization of manufacturing technics and the assurance of product quality as objectives, virtual technology in manufacturing process implement production process planning, organization management, shop scheduling, supply chain and modeling and simulation of logistics design.

## 3.  Objectives and Tasks

With the deep development of artificial neural network, expert system, fuzzy logic and genetic algorithm, network-based manufacturing technology in production and operation of enterprise, the meaning of integration and intelligentization is deepening. System integrated manufacturing uses intelligent activities of human experts simulated by computer to carry out analysis, judge, reasoning, thinking and decision-making so that experience's decision-making and management can transfer to scientific decision-making and management supported by intelligent information management. The boundaries between machining and design are fading out and will move towards integration. Enterprise process control will integrate with production and operation, and management and control for production and operation process will be achieved with self-tuning, adaptive, optimal, fuzzy control, etc. The boundaries among machining process, monitoring process, logistics process and assembly process are fading out and disappearing and they will be integrated into a unified manufacturing system. Information integration, functional integration and process integration in manufacturing system will achieve products' business collaboration, product design collaboration, resource sharing, manufacturing collaboration and supply chain collaboration. Furthermore, digital and intelligent products, as well as digital and intelligent product design, development, production, testing, management and maintenance process should be achieved.

Chinese manufacturing industries continuously adopt advanced manufacturing technology, but compared with industrial developed countries, there are still some gaps. The level of information, automation, intelligentization of manufacturing management and process control system of enterprise are not high. At present, the applications of information-based technology of domestic enterprises are mainly in the phases of unit technology applications and enterprise integration application, and enterprises will be in these two phases for long time. Integrated application in Internet environment is still in the exploratory stage and only a small number of enterprises gave demonstration. Informationization is the basis and premise of intelligent manufacturing, and is the significant mean of promoting and reforming traditional industries, and is the tool of revitalizing manufacturing industry.

It is believed that, based on informationization, manufacturing system will develop towards intelligentization, integration, networking, virtualizationand coordination. Popularization and application of information technology can promote technical innovation and management innovation of enterprise. Digital product design, network-based enterprise management, automated production processes, intelligent production equipment will be realized. Integrated manufacturing system will have the ability of amending or reconstructing its structure and parameters, and have elf-organization ability and coordination ability. The initiative and flexibility of system will be demonstrated.

## 4. Development Roadmap

System integrated manufacturing is to integrate automation, integration, network and intelligence together. It uses intelligent activities of human experts simulated by computer to consider system and related requirements with a kind of integrated method. Changing market demand and fierce competitive environment are met by dealing with massive and complex information. It must become the main production mode of future manufacturing industry.

System integrated manufacturing should further deepen and integrate traditional manufacturing system firstly. The integration of control hardware, software, intelligent information processing method should be paid more attention. Secondly, modern management and design, computer and system science, artificial intelligence should be organically integrated to provide new ideas, new methods and new technologies for intelligent manufacturing, and to impel its development towards to a higher level.

By 2020, it is predicted that manufacturing system with information integration as core will be applied widely, and the productivity will be increased by over 10%. Innovative product design methods, e.g. product lifecycle management (PLM) technology, multi-disciplinary design optimization technology, will change traditional product design concepts and approaches. Manufacturing system modeling, simulation and optimization technology, distributed artificial intelligence, large-scale systems decision-making, distributed network system mode based on multi-intelligent agents will be developed. New manufacturing modes such as agile manufacturing, human-machine integration, etc. will change traditional concepts of production organization mode of manufacturing industry. Intelligent planning, execution and scheduling system, collaborative product business technology, modern logistics technology and virtual manufacturing will have new breakthrough.

By 2030, knowledge management platform, advanced manufacturing system integration platform based on network, enterprise collaboration technology will provide cross-region network operating environment for enterprises' design, production, management and marketing, etc, and enable enterprises to move towards globalization, integration and ordering. Distributed intelligent operation model and performance analysis technology, massive data processing of manufacturing information and manufacturing knowledge database management technology, negotiation and mediation technology among hierarchical intelligent agents, perception and learning mechanism of intelligent agents in distributed manufacturing, interactive dynamic knowledge acquisition and knowledge automatic acquisition will be solved. Rapid response manufacturing technology, key technics and integrated technology, manufacturing technology of special manufacturing equipments, manufacturing technology of new/special materials and intelligent manufacturing plan and execution system will enable manufacturing resource to achieve sharing and optimal utilization. Cross-region and cross-country network-based collaborative manufacturing system will be established initially.

By 2050, the intelligentization will be widely applied in the entire process of manufacturing system. The intelligent system will be able to change its behavior according to environment or objectives, the controlled objects will have the ability of 'learning' and 'recognition', the system will be adaptive and robust to the changes of environment and disturbance, and the ability of manufacturing enterprise, system and unit adapting to environment will be greatly improved. Intelligent design methods and tools,distributed intelligent resources generation technology use computer to simulate thinking activities of human and increase intelligent level of design. Intelligent resource planning, which searches optimal solution and rapidly execute according to objectives set, will be widely applied in the process of production business. Mechanical and life sciences information science, materials science will be highly integrated, and bionic control theory system, design system based on image recognition of product modeling, growth and forming process, bionic design and manufacturing, intelligent bionic machine and bio-forming manufacturing and so on will have breakthroughs, and virtual, convergent and intelligent manufacturing enterprise system will be achieved.

According to the basis status, development trends and possible technical breakthroughs of existing technologies, development roadmap of aforementioned relating technologies in different phases by 2050 is given.

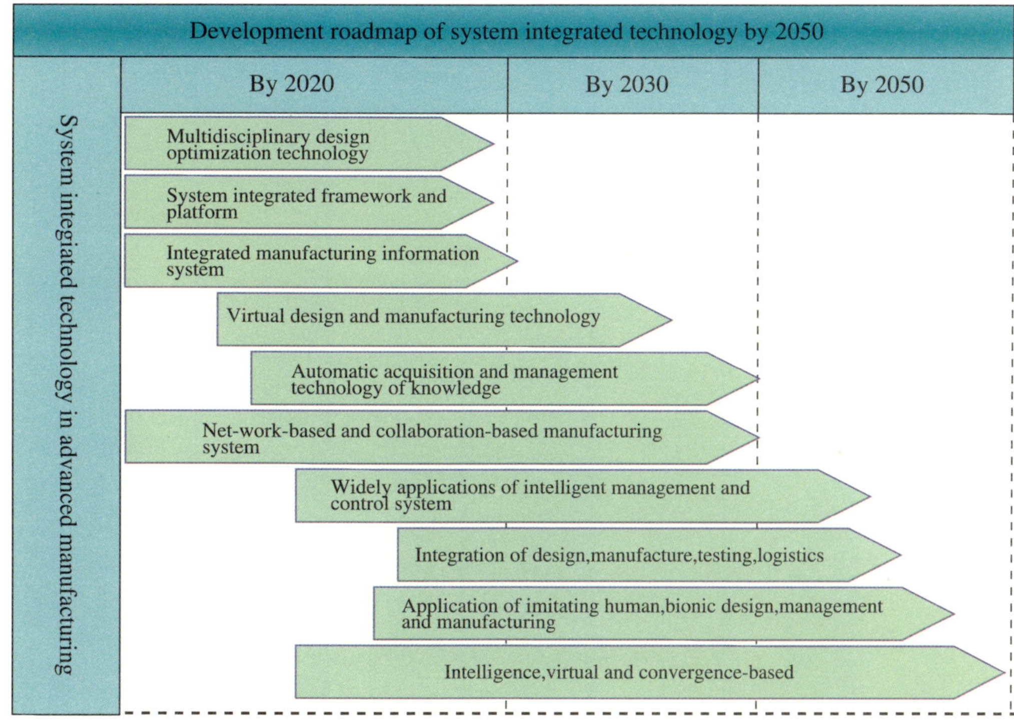


Figure 3.9    Development roadmap of system integrated technology

## 5. Summary

System integrated manufacturing combines automatization, integration, and intelligentization in one, is a kind of advanced manufacturing system characterized by continuously developing towards high technology content and high technology level. It is also a kind of human-machine integrated system composed by intelligent machines and human experts together. It emphasizes that, in all aspects of manufacturing, analysis, judge, reasoning, thinking and decision-making should be achieved in a highly flexible and integrated form, through the intelligent activities of human experts simulated by computer, and intelligent manufacturing of human experts will be inherited and improved.

# References


[1] Bureaux of significant projects management for information-based engineering of manufacturing industry. 2002. Training material of information-based engineering of manufacturing industry.

[2] Cui Y. Perspective of manufacturing industry on the era of globalization: Seeking win-win in strong interaction. http://cpc.people.com.cn/GB/64093/82429/83083/7164596.html.

[3] Engels F. 1985. Dialectics of Nature. Changchun: Jilin University Press.

[4] Experts group in advanced manufacturing technology field. 2007. '11th Five-Year Plan' Development Strategy Report of Advanced Manufacturing Technology Field in National 863 Program.

[5] Fan Y S. 2005. Enterprise innovation supported by information technology. 11th China (Tianjin) Information Technology Exposition - International Exchange Conference of IT Personnel 2005 National Information Development Forum.

[6] IBM global information technology service department. CCW Research. 2008. White paper of developing trends of Chinese information development.

[7] Lead group of information-based engineering of Chinese manufacturing industry. 2003. Developing Report of Information-based Chinese Manufacturing Industry. Beijing: China Machine Press.

[8] Li Z, Qu X M. 2008. Developing Report of Chinese Equipment Manufacturing Industry in 2006-2007. Beijing: China Planning Press

[9] Modern History of Mechanical Engineering. http://www.ikepu.com/machinery/machinery_history/contemporary_history_mechanical_total.htm.

[10] Office of the Secretary of Defense. 2005. Unmanned Aircraft Systems Roadmap:2005-2030. http://fas.org/irp/program/collect/uav_roadmap2005.pdf.

[11] Qi G N. 2008, Some ideas about developing trends of advanced manufacturing technology. 2th Phoenix Contact Annual Academic Meeting.

[12] Research group of technology foresight of China towards 2020. 2008. Technology Foresight of China Towards 2020. Beijing: Science Press.

[13] Synthetic special group of China Association for Science and Technology and developing research of manufacturing technology. 2004. Developing research of Chinese manufacturing technology in 2020.

[14] Yang S Z, Wu B. 2003. Trends in the development of advanced manufacturing technology. Chinese Journal of Mechanical Engineering, 39(10): 73-78.

[15] Yang S Z, Wu B, Li B. 2006. Further discussion on trends in the development of advanced manufacturing technology. Chinese Journal of Mechanical Engineering, 42(1): 1-4.

# Part II

## Environment Friendly Green Manufacturing

# 1 Overview

Global manufacturing industry is experiencing a profound evolution which is characterized as high value, high technology, and information-based. At the same time, sustainable green manufacturing - low carbon economy - becomes a main trend and a human-common strategic objective. According to the domestic situations, entity manufacturing industry is still dominating our national economy in the long term. As one of the largest countries in the global manufacturing industry, most of our products are, however, still in the low end of the international industrial chain, providing the basic raw materials, production and consumption products to the world. The low resource utilization rate and productivity where a large gap still exists compared with the developed countries, makes our resources and environment burdens extremely heavy and becomes a bottleneck of the sustainable development of the economy. Intensifying the green-based technologies research and development in the manufacturing industry, therefore, becomes particularly urgent.

The development strategic frameworks for the green manufacturing technologies in China by 2050 are as follows:

(1) Increase the green-based technology upgrading in process manufacturing for large-scale resources conversion and utilization of the minerals, oil and gas, and biomass. Innovative breakthroughs are expected to be achieved in the green process and engineering of the materials conversion. In the main research fields, e.g. new reaction medium substitution technology, high efficiency catalyst technology, process intensification technology, advanced reaction separation equipment, material recycling, and critical environmental technology, we should develop green process technologies which can significantly increase the resources utilization rate and dramatically decrease the energy consumption and waste discharge. By 2050, raw material loss rate in the process manufacturing industry will be reduced by 90% and the atom economy and selectivity in the process will reach 100%. In addition, zero waste emission will be achieved and chemical environmental risks will be eliminated. Consequently, a green-based manufacturing industrial system can be created.

(2) New raw material, new resources substitution technologies in

the process manufacturing industry and biomass processing and refining engineering should achieve innovative breakthroughs. Till 2050, cost-effective separation and recycling of C1 compounds and $CO_2$ will be greatly increased. Biological-based chemical and biological materials will account for 45% of the total resource share. Unconventional mineral resources will be partially used, and a low-carbon economic industrial mode will be constructed.

(3) Product green design and full life-cycle assessment system in the discrete manufacturing industry will be applied extensively. Scrapless machining technology as well as  recycling and remanufacturing technology of electromechanical products and automobiles will be popularized, forming an arteries and veins integrated production system. By 2050, global recycling social system and advanced technology system based on low carbon economic mode will be created, which will make China to be one of the advanced countries in the world.

The construction features and objectives of green manufacturing industrial system in China by 2050 are listed in table 1.1。

**Table 1.1　The construction features and objectives of green manufacturing industrial system in China by 2050**

| | | Around2020 | Around2020 | Around2020 |
|---|---|---|---|---|
| Green process | Saving energy and reducing carbon emissions | Energy will be saved by 30%,and carbon emission will be reduced by 20% in manufacturing process | Energy will be saved by 50%,and carbon emission will be reduced by 50% | Low carbon economy manufacturing industry system will be established |
| | High efficiency and clean utilization of resources | Loss rate of raw material will be reduced by 30%,and recycling utilization rate of secondary resources will come to 50% | Loss rate of raw material will be reduced by 50%,and recycling utilization rate of secondary resources will come to 70% | Loss rate of raw material will be reduced by 90%,and recycling utilization rate of secondary resources will come to 90% |
| | Environmental impact | The environmental pollution in manufacturing process will be basically controlled | The hazardous waste will approximately achieve zero discharge,and chemical environmental risks will be basically cotrolled | Zero discharge of waste will be achieved and chemical environmental risks will be basically eliminated |
| | Products green design | Easy of disassembly and recovery of mechanical and electrical products and automobiles will be achieved | Green design standards for the full cycle of major products will be establised | Full cycle green design and recycling utilization of products will be popularized |

 *Advanced Manufacturing Technology in China: A Roadmap to 2050*

# 2 Field Features

## 2.1 Green-based Upgrading Demands of the Domestic Manufacturing Industry

China is one of the largest manufacturing countries in the world and the total output value from the manufacturing industry accounts for 40% of the national GDP, making it the core of the national economy. It makes us the 4th largest manufacturing country with the output ranking No. 1 of more than 80 products in the world. However, the labor productivity of manufacturing industry is low and most of the products are easily consumable and low value-added, locating on the low-end of the international labor division, and many products stay in the stage of simple processing which restricts the market expansion only by products scaling-up. The developing level of our equipment manufacturing industry, which represents the technology level of the manufacturing industry, is far behind the developed countries. Most of the manufacturing enterprises are lack of core technologies, which leads to excess dependency on the international capitals and foreign technologies. Particularly, the technology level of the domestic manufacturing industry is far behind the international advanced level. In China, the resources utilization is featured by low resource utilization rate and low productivity; energy consumption and industrial wastes produced in the domestic industries induce serious environmental pollution and severe waste of resources. For example, the unit energy consumption of the main industrial products for the domestic enterprises is 20%~70% higher than that of the international advanced level (Japan); each year, hundreds of million tons of industrial wastes are discharged from metallurgical and chemical industries etc. which cause serious environmental pollutions; the discharge amount of waste water per GDP in China is 4 times higher than that of the developed countries and the solid waste is 10 times higher. The problems of the resources and the environment in the manufacturing industry constrain the sustainable development of the national economy. According to the research results of the National mid-long term technology strategy planning, the effect of the resources utilization on the ecology in 2020 will be 4 times more if we keep the current energy utilization rate and waste discharge level, and the sustainable development will be even more difficult.

Resources reserves per capita in China are low and the per capita share of 45 major mineral resources is lower than half of the world average. With the fast development of heavy chemical industry, the dependency on foreign countries for the main mineral resources such as oil, gas, iron, aluminum, and copper etc. will increase and the sustainable development rate will decrease dramatically.

In the next 30 to 50 years, the population in China will still increase steadily, and the requirement of society development for energy resources and materials will keep increasing. At the mean time, the manufacturing industry will dominate the national economy in the long term. In order to become an advanced manufacturing country, we need to depend on the scientific and technical innovation to solve the bottleneck problems of the resources and environment sustainable development from the beginning. The extensive mode of economical growth should be got rid of and green-based upgrading of manufacturing industry should be accelerated. Resource saving and environment friendly green manufacturing technology should be adopted, alternating the economical growth mode by means of the scientific and technical advancement.

With the global environmental degradation and resource shortage, green-based manufacturing industry has become an urgent request of the human society's sustainable development, and hence the conception of green manufacturing generates. Green Manufacturing is a sustainable manufacturing mode for the integrated resources utilization and environmental effect. Its objectives are to minimize the influence of environment and maximize the utilization of resources in the whole product life cycle-from design, manufacturing, packaging, transportation, and utilization to scrap. With the control of reducing source and resource recycling, enterprise economic benefits and society environmental benefits will be optimized coordinately. The system structure includes green design, material selection orienting environment, green techniques, green package, recovery and re-manufacturing technology, bio-industry technology orienting environment and resources, evaluation and decision-making analysis in the full life cycle. It covers both the discrete manufacturing industry and the process manufacturing industry. Nowadays, the green manufacturing has been integrated with clean production, resource recycling and utilization, and high value-added products, forming the development strategy of the manufacturing industry.

In order to achieve the sustainable development and develop green manufacturing technologies, the manufacturing industries are facing the following problems and trends:

(1) Urgent requirements to dramatically increase the utilization efficiency of energy and resources exist to develop low waste emission clean production and advanced cyclic economical technologies.

(2) New clean, high efficiency, and recycling technologies, and processes are required to be developed urgently for our local resources, secondary resources, bio-resources, and other alternative resources, which will provide resources security guarantee for the sustainable development in the next 30~50 years.

(3) We should take the opportunities of manufacturing transformation, and lead the scientific and technical development to reconstruct the domestic industry system. The basic raw materials processing should be highly integrated with the high value-added products service, and the same for the discrete manufacturing industry and the process manufacturing industry. Consequently, a domestic green manufacturing industrial structure and advanced technology system will be established, making China to be one of the advanced countries in the world in 50 years.

## 2.2 Development Trends of Green Manufacturing Technology

"Green Plans" in the theme of the environment protection have been proposed by many countries e.g., in Europe, Japan, US, and Canada from 1990s in order to promote the R & D of the green manufacturing technologies, legislations, environmental standards, objectives, and strategies for enterprises. At the beginning of the 21st century, the US proposed that the use of raw materials and comprehensive consumption of energy resources and waste discharge should be reduced by 30% in the next 20 years, and the related green technologies should be developed; the 2025 plan of Japan proposed that environment-oriented and technology strategic chemical technologies should be highly developed, which lies in environment-oriented alternative new technology, new process, and raw materials substitute, creating a pollution prevention and resources cyclic system (including biology ocean resources). Recently, with the increasingly severe global climate change effect, low carbon economic technologies including low carbon products and techniques will be paid much more attention by developed countries gradually. Therefore, carbon-intensive production processes and modes with high energy and materials consumption will be changed dramatically.

Manufacturing industries in the developed countries have finished the transformation from the mass primary products to high value-added products production and from simple manufacturing to product whole life service. The difference between manufacturing and service blurs, and the main income of the manufacturing industry will not only come from the primary products but focus on the high value manufacturing, which is a complicated network including design, research and development, production, sales, logistics, whole life service, and products life cycle. Accordingly, UK has established the UK Technology Strategy Board, which issued the 2008~2010 development strategy of the high value manufacturing critical technologies. It realizes the above changes profoundly and confirms the technical theme that the UK manufacturing companies will transfer towards high value activities.

With the rapid development of the information technologies, industrial economy is turning to knowledge economy, and entity manufacturing is turning to "virtual manufacturing". As well, complete manufacturing is turning to network-based assembly manufacturing. As a result, the cost of the non-entity manufacturing process, e.g. research, development, design, information processing, and model test, will be higher than the cost for entity-based manufacturing process gradually. From space distribution, the vertical specialization change of the assistant system will enable the whole production process of the manufacturing industry to scatter around the world to achieve the optimal resources allocation. This causes that the technologies and knowledge can exchange more easily in a relatively centralized range and promotes innovation, and therefore the upgrade from manufacturing to creation will be achieved. The above transformations are gradually blurring the boundary of the discrete manufacturing industry and the process manufacturing industry. From the long-term perspective, the discrete manufacturing industry will develop towards network-based assembly manufacturing and the process manufacturing industry will develop towards the integration of resources-based and discrete-based products production.

The oil, gas, and mineral are the basis of the energy resources and materials supporting the development of the global manufacturing industry, but a severe challenge presents of energy resources shortage and resources exhaustion. The wide utilization of new energy resources such as renewable energy resource and nuclear power will change the current energy resources structure of the manufacturing industry; alternative resources, e.g. low quality resources, bio-mass resources, and secondary resources will dominate manufacturing industry resources processing gradually. Meanwhile, the demands of resource recycling will also change the current structure and utilization methods of the products. From the beginning of the 21st century, knowledge spread and update speed have been much faster, and the technical breakthroughs of material, life, and information fields in the transformations and applications of manufacturing industry field are also faster and faster. The profound integration of high technology and manufacturing industry will introduce high technology-based into the traditional manufacturing industry. Advanced control, bio-process, and nanotechnology have become the sources of innovation technologies of the manufacturing industries in developed countries, and green manufacturing has already become the main direction of the manufacturing scientific and technical development. These will enhance the alternation of production technologies and methods in the manufacturing industry. Meanwhile, biology-based industry has become a preferred direction of the manufacturing industry in developed countries in order to increase the resources utilization efficiency.

In summary, manufacturing industries in developed countries such as the US, Japan, and Europe etc. are still the main body of the national economy, and will occupy the high-end of the value chain in the global manufacturing industry and lead the technical progress. Therefore, the whole industrial chain will be controlled by reforming the product mode and establishing new standards.

## 2.3　Scientific Connotation and Methods of Green Manufacturing

Green manufacturing is a comprehensive process system which uses atom economy conception of material conversion and the principles of symbiotic species, and integrates the high-efficiency hierarchical multi-level utilization of material designed according to the system engineering and the optimized method to make full use of materials and reduce waste emission from the beginning aiming to solve the environmental and recourse problems. It includes green design and process intensification of atom economy reaction and separation process during the efficient and clean conversion of materials from micro scale, as well as the process coupling and adjustment from global scale, in addition to the optimal integration and environment-based multi-objective optimization of material flow-energy flow-information flow and system integration of biology industrial group of global system level. A full-optimization will be achieved to provide support for the sustainable development of the manufacturing industry.

The scientific connotation and methods of green manufacturing in the process manufacturing industry and the discrete manufacturing industry are shown in Figure 2.1 and Figure 2.2.

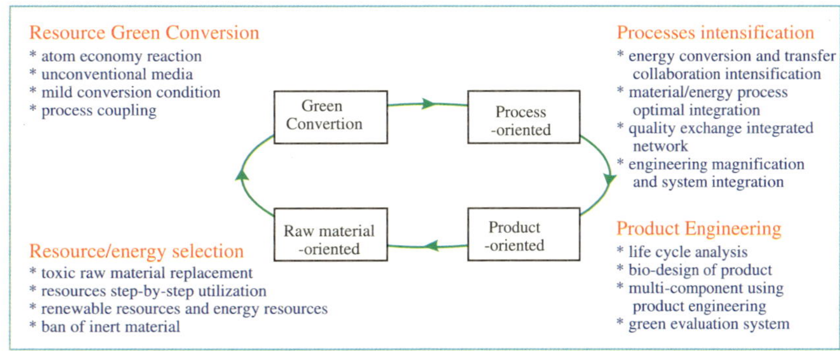


Figure 2.1　The Scientific connotation and method of green manufacturing in process manufacturing industry

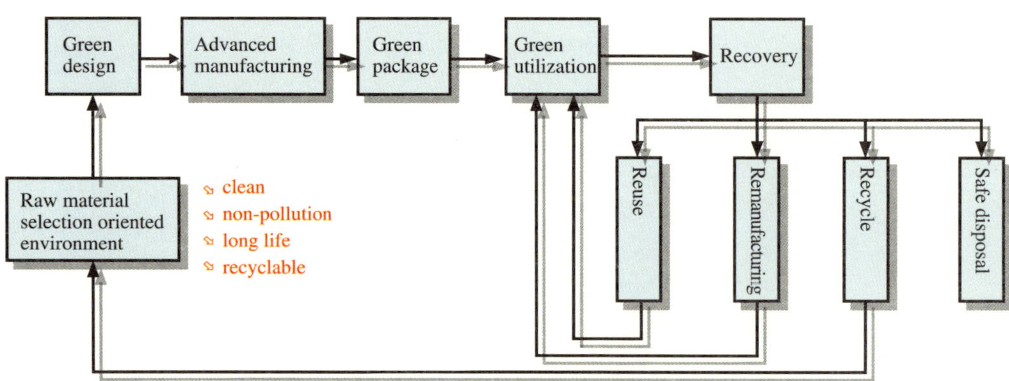


Figure 2.2    The Scientific connotation and method of green manufacturing in equipment manufacturing industry

# 3 Development Objectives and Key Technologies of Green Manufacturing Innovation System

## 3.1 Development Objectives and Strategic Frameworks of Domestic Green Manufacturing Innovation System

In order to achieve the promotion of the domestic green-based process industry, the development objectives and strategic frameworks of the domestic green manufacturing technology by 2050 are as follows:

(1) The upgrades of green-based technologies in process manufacturing for large-scale mine, oil and gas, and biomass resources conversion and utilization should fulfill innovative breakthroughs in the green process and engineering of material conversion and should establish green process engineering technologies in the new reaction medium substitution technology, efficient catalyst technology, process intensification technology and advanced reaction separation equipment and recycling of secondary resources and environmental core technology, which can greatly increase utilization rate of resources and reduce energy consumption and waste discharge. By 2020, material loss ratio of process manufacturing industry will reduce by 30%; recycling utilization of secondary resources will come to 50%; process energy saving will come to 30%; carbon discharge will  decrease by 20%; and environmental pollution of manufacturing process will be basically controlled. By 2030, material loss rate of process manufacturing industry will reduce by 50%; recycling utilization of secondary resources will come to 70%; process energy saving will come to 50%; carbon discharge will decrease by 50%; and the discharge of hazardous waste will approach zero and chemical environmental risks will be basically controlled. By 2050, material loss rate of process manufacturing industry will reduce by 90% and atom economy and selectivity

in chemical conversion will approach 100%. Zero discharge of waste will be achieved and chemical environmental risks will be eliminated. Green-based manufacturing industry will be established.

(2) Innovative breakthroughs will be acquired in new material and resource substitution technologies in the manufacturing industry and biomass processing refinery engineering. By 2020, energy consumptions of straw pre-processing will be lower at least 50%, the costs of cellulase will decrease by 4~5 times. By 2025, bio-ethylene will be obtained. By 2035, large-scale production and application of the bio-propylene refinery technology will be achieved. By 2050, application of low-cost application separation and conversion for C1 compounds and will be intensively improved. Bio-based chemicals and biological materials will account for 45% of total share. Unconventional mineral resources will be applied partially, and low carbon economic mode will be constructed.

(3) Green design and full life cycle assessment system in discrete manufacturing industry will be applied widely. The scrapless machining technology, the recycling and remanufacturing technology of electromechanical products and automobiles will be popularized, forming an arteries and veins integrated production system.

(4) By 2050, the advanced technology system of constructing global recycling social system and low carbon economic mode will be set up. China will be in the rank of advanced countries in the world.

For a green process engineering system with efficient, clean, and recycling resource utilization, the core scientific and technical issues are: firstly material conversion, multi-scale mechanisms of recycling, adjusting and control methods, and principles of engineering optimized amplification for efficient and clean resource utilization should be revealed. Core technologies of green process engineering should be broken through, and new techniques, process, equipment, and integrated technologies should be established. Secondly product green design and new full life cycle assessment methods should be established, and the resource recycling and environment control technology, product disassemble and easy reuse technology, low-cost separation and resource-based utilization technology for $CO_2$, etc. should be broken through. Thirdly the multi-scale design, engineering demonstration, and technology integration should be carried out and an arteries and veins integrated production system should be developed.

In the next 20 years, material loss of process industry will reduce by 50%; energy saving of unit product will come to 30%~50%; discharge of hazardous waste will approach zero; chemical environmental risks of industrial manufacturing process will be basically controlled; recycling utilization of secondary resources will come to more than 60%. The pollutants controlling from the beginning adapting to the domestic status, green process engineering technical system of material recycling utilization will be established and be applied in industry.

The node objectives and roadmap of scientific and technological development in the green manufacturing field by 2050 are given as follows.

    *Advanced Manufacturing Technology in China: A Roadmap to 2050*

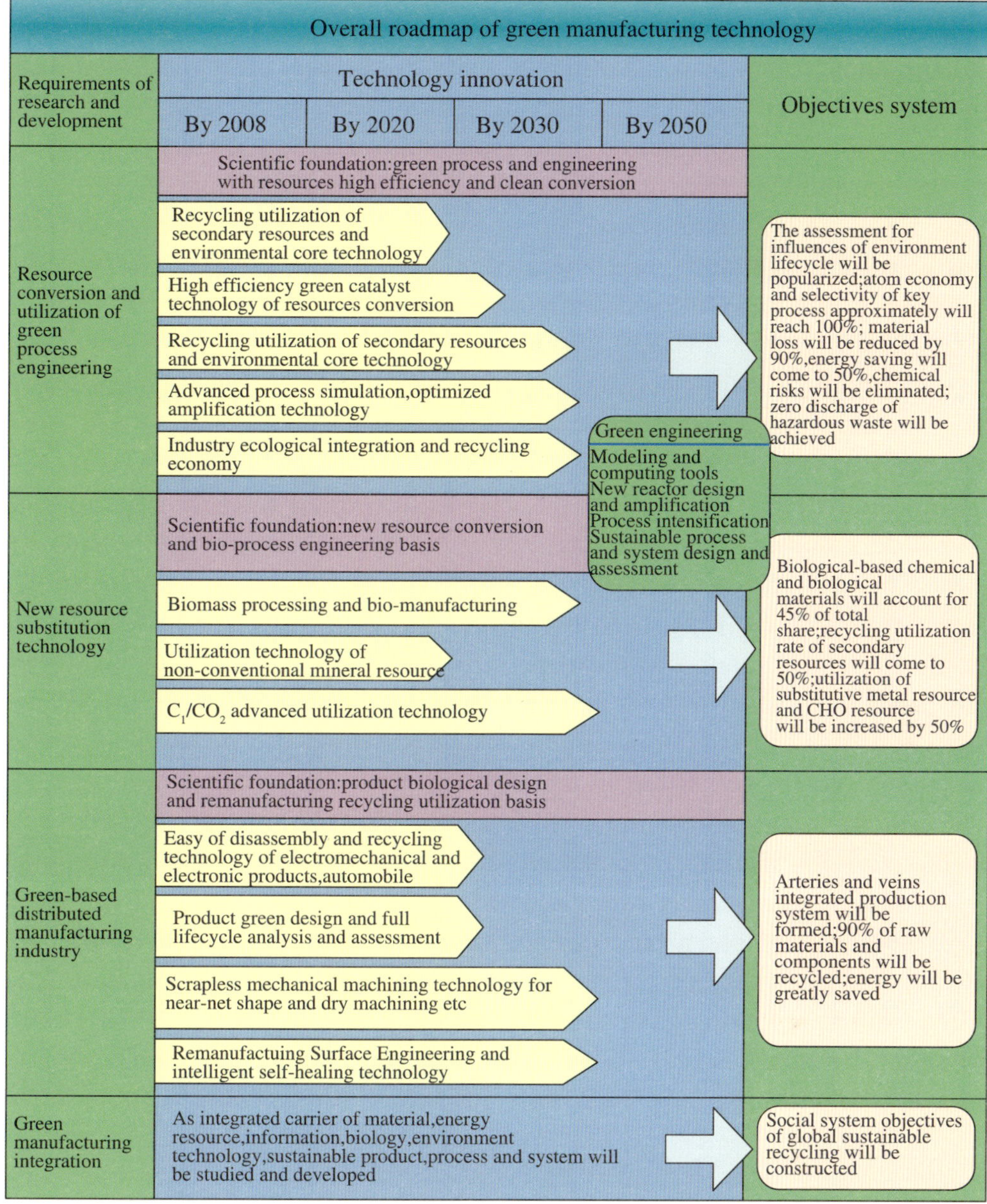



# 3.2  Green Process Engineering of Mineral Resource Utilization

## 1. Demands and Challenges

Mineral resources are important material foundations to ensure the economic development and national security. In the stage of rapid development of the domestic industrialization, the contradiction between the supply and demand of mineral resources is emerging and the storage of mineral resources to be used in industry is very deficient. The overall utilization rate of resources is very low along with serious environmental pollutions, and the total consumption of energy resources increases dramatically. This severely limits the sustainable and coordinated development of the domestic national economy and threatens the mineral resource security of national strategy. Based on the domestic mineral resource features and traditional extraction technologies, efficient and clean processes of separation and extraction should be developed in the aim of the insufficient strategic mineral resources utilization. At the same time, we should enlarge the amount of the strategic mineral resources used in industry and deal with the bottleneck of resources and environmental problems in minerals and metallurgical industries which is the strategic requirements of China.

With the continuously enlargement of minerals mining, processing, and utilization, each country is facing the problems of resource and energy shortage, environmental degradation. Thus, much more attention has been paid to the efficient-clean comprehensive utilization of mineral resources. The developed countries such as the United States, Japan, and Europe, have list the related researches into their schedules of national strategic high technology development, and therefore fundamental, applied, and industrialized researches of efficient and clean production technologies have been given enormous supports. Meanwhile, the related laws and corresponding technical specifications and rules have been issued to impel the mining solid waste treatment and the development of the integrated utilization technologies. The 21st century strategies have been proposed of developing pollution-free green products which will focus on the technical researches of efficient extraction-separation and comprehensive utilization of mineral resources.

The current research trends of resource biological utilization in the aim of resource efficient conversion are as follows: efficient conversion and utilization of resource; clean utilization and conversion and environmental friendly technologies of resources; optimized integration of material – energy – information flow in the whole process.

## 2. Objectives and Tasks

Due to the increasingly serious resource and environmental problems, the future development of mineral resources conversion processing technologies will be aiming at efficient and clean extraction and recycling of mineral resources. In the main research fields including new reaction medium substitution technology, high efficiency catalyst technology, process intensification technology, advanced reaction separation equipment, material recycling, and critical environmental technology, we should develop green process technologies which can significantly increase the resources utilization rate and dramatically decrease the energy consumption and waste discharge. A series of common technical innovations for efficient, clean, and cost-effective resource conversion-production, in virtue of process intensification and system integration among multi-unit/multi-process will be developed to achieve the objectives of maximum resource utilization, minimal energy consumption and environmental load.

By 2050, the overall objectives are: critical innovative technologies will be developed to achieve nearly 100% of the atom economy and selectivity in the critical parts of minerals processing; materials loss will be reduced by 90%; energy saving will reach more than 50%; and zero waste discharge will be fulfilled and chemical risks will be eliminated.

## 3. Critical Technologies

In recent decades, resource conversion processing technologies have attained rapid development and fundamental researches have been extended in both depth and breadth. The differentiation and intersection among disciplines have been enhanced and the relationships between science and technology become much closer. In the next 50 years, in order to develop new utilization technologies of mineral resources, breakthroughs in the following fields need to be achieved:

### 1) High efficiency conversion technology of unconventional reaction mediums for mineral resources

Unconventional reaction mediums with excellent physical and chemical characteristics should be developed to break down the thermodynamics and kinetic constraints of the conventional mediums for existing mineral resources. The critical common technologies of unconventional medium activated M-O and M-S bond intensified conversion for mineral resources will be generated. For example, the common technologies by using the unconventional reaction medium of sub-molten salt to actively decompose the multiple amphoteric metal mineral resources such as Cr, Al, Ti, V, Mn, and etc., and using the unconventional medium of ionic liquid low-temperature electrolysis to produce reactive metals such as Al, Mg, and etc. can increase the resource utilization rate in the existing process and reduce the reaction temperatures and energy consumption dramatically. By 2030, the comprehensive energy consumption of the mineral resources conversion will be reduced by 40% and the material loss will be decreased by 80%, approaching zero waste emission.

### 2) Integrated technologies of resources and products

Advanced and integrated technologies of resources processing and machining should be developed. The products can be directly made from minerals by physical and chemical processing, and accurate controlling of structures and properties of the materials should be achieved. By improving the performances machining properties of difficult-machining materials, the seamless links between process manufacturing and functional product manufacturing process can be achieved. For instance, the processes of the pyrometallurgical and hydrometallurgical technologies of high value-added materials, functional material preparation, and coupling hydrometallurgical technologies of metal-based compounds and sulfur production will be shortened dramatically and the green degree can be increased.

### 3) Process intensification technologies of mineral resources conversion utilization

Process intensification includes production equipment intensification and production technique intensification, such as new reactors with intensive dynamics, new heat transfer and separation equipments, new technical conditions, operation of the intensive dynamics process, and etc.. They are directly related to the green degree of the process. In order to reach low pollution, low energy consumption, and short flow in the pyrometallurgical melting processes, imperial smelting furnace, flash smelting, and bath smelting technologies have already appeared and the melting of complex minerals become easier. It is predicted that new breakthroughs will be achieved by controlling oxygen flow.

Much more attention has been paid on the hydrometallurgical technologies due to their resources and environmental advantages during the processing of low-grade, refractory, and complex multi-element minerals in the recent several decades. Some pollution-free hydrometallurgical technologies, e.g. pressure hydrometallurgy, unconventional reaction medium hydrometallurgy, biological hydrometallurgy, supercritical fluid technology, liquid membrane separation technology, and new extraction and separation system, have achieved significant progress. In the intensive dynamics and separation aspects, the hydrometallurgical technologies are expected to be in the world frontier.

Combing with the computer controlling techniques, the electrolysis technologies have been improved gradually. The size, current, and energy efficiencies of the electrobath have been greatly increased. The several present new electrochemical methods used to extract metals will induce important application achievements, such as direct electrolysis method of oxides, solid electrolyte-based medium electrolysis to obtain metals, electrochemical deposition method with ionic liquid and other cutting edge electrochemical technologies.

### 4) In-situ leaching and bio-leaching of orebody technologies

Low energy consumption, cost-effective hydrometallurgy technologies such as in situ leaching and bio-leaching of ore body are developed rapidly. This technology leaches the materials in-situ without mining, and without waste gases, waste water, and waste ores. Therefore, it will not generate any environmental pollution, deforestation, and ecological destruction, and avoid the energy consumption during transportation. It is suitable for ore bodies especially for the domestic ore bodies which are low-grade, deeply buried, difficult mining, and uneconomical using conventional mining technologies.

### 5) High efficient utilization of marine mineral resources

Due to the increasingly shortage of the land mineral resources, minerals mining will gradually turn to the oceans with rich mineral resources. China is a big costal country with abundant marine mineral recourses. The new cost-effective technologies should be developed that can effectively separate and extract the valuable elements from unconventional mineral resources/ substitution resources, e.g. marine mineral recourses (manganese nodule, cobalt nodule, metal in seawater, hydrothermal sulfide deposit in the seafloor), salt lake and metal salts in lakes, heavy metal sludge.

### 6) Novel green-based upgrading process for iron and steel production

(1) Smelting reduction technology: blast furnace ironmaking technology is facing such problems as the shortage of coke and coal, long production process, and environmental pollution caused during coke production, and therefore the present ironmaking technology is transforming towards the directions of low coke or coke-free. It mainly includes direct reduction and smelting reduction technology and other non-blast furnace ironmaking technologies. In the smelting reduction technology, coke is replaced by coal and iron ore powder can be directly used during the iron making by avoiding the coking, sintering or pelletizing processes. As a result, green-based upgrading technologies are expected to be achieved in iron and steel making process. In the next 50 years, smelting reduction technology is expected to be applied extensively.

(2) Hydrogen metallurgy technology: the large-scale and cheap hydrogen preparation technology may be broken through in the 21st century. Hydrogen metallurgy is a technology where hydrogen is used as reductant instead of carbon. The reducing potential of hydrogen is far higher than carbon, resulting in high reduction rate and efficiency, and consequently, great reduction of carbon consumption. This guarantees the sustainable development of iron and steel making industry. The high purity iron reduced by hydrogen can promote the new technical revolution of steel making, continuous casting, and steel rolling techniques. The new process of iron and steel making in the 21st century will be formed and a new generation of iron and steel materials with high purity, toughness, and corrosion resistance will be developed.

(3) Advanced pretreatment technology for refractory low grade iron ore. For example, the energy consumption in recycling of low grade iron resources can be saved by 50% by replacing the conventional vertical furnaces with advanced fluidized beds and magnetized roasting pretreatment in a rotary kiln.

### 7) Intelligent control for resource machining process

Advanced models and information technologies will be widely used to construct solid machining engineering database of resource conversion. They combine with the computer technology and advanced control technology to achieve intelligence-based machining process in resource conversion. If the computer numerical modeling and process simulation technology are developed to design and optimize the machining methods and techniques, the accurate control in resource conversion process can be achieved. By 2030, it is expected that on-line monitoring and intelligent feedback control in resource conversion process will be achieved, which will promote the production reliability, stability, and efficiency significantly. Meanwhile, the material consumption and waste discharge will be reduced effectively. Reasonable combination and highly coordinated optimization of the process system will be obtained. Technological structure will be adjusted and optimized. The economized production process, clean-based smelting processes, and product diversification and intensification of flexible production capacity will be achieved.

## 4. Development Roadmap

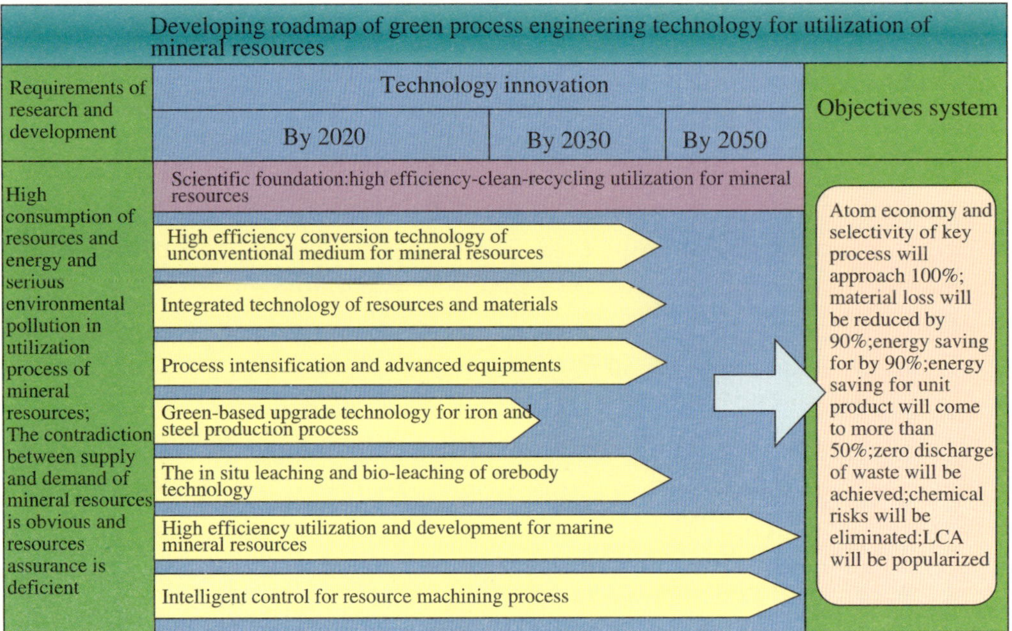


## 5. Summary

In order to solve the bottleneck problems caused by resources, energy, and environment during processing and utilization of mineral resources and the deficiency of security with respect to  industry resources, high efficient-clean-recycling mineral resource utilization should be achieved with the help of technological breakthroughs and industrial applications of the green-based upgrading technology for iron and steel making process, efficient conversion technology of unconventional medium for mineral resources, process intensification technology of resource utilization, and the integrated technology of resources and materials. The ore in-situ leaching and bio-leaching technology and efficient exploration and utilization of marine resources can ease the shortage of mineral resource deficiency. Intelligent control for resource processing is the generic key technology of mineral resource utilization. The execution of green process engineering technologies mentioned above for mineral resource utilization provides technical supports for sustainable development of the domestic mining and metallurgical industries.

Conversion and machining process of mineral resources: it uses original and secondary resources and varieties of chemical/physical/biological methods to extract the valuable components. After they are processed and converted, they will become metal or nonmetal material which possess certain utilization performance and are taken as integrated machining process of products. The progress and development of these technologies are the premise and foundation of industrial development of modern metal materials. Considering these technologies as core, process manufacturing industry is an important pillar industry of national economy, and also is pillar industry of other national economy. For example, mechanical manufacturing, automobile, petrochemical, transportation, and so on provide important material foundation.

Sub-molten salt unconventional media: it is multi-salt flow system whose water content is less than 50%. It is characterized by low vapor pressure, high boiling point, good fluidity, strong chemical activity, performance controllability, and etc. Reaction system of low temperature unconventional medium of Sub-molten salt has the advantages of thermodynamic and kinetic properties. It greatly enhances reaction process and achieves high efficiency decomposition and conversion of mineral under mild conditions.

Underground leaching: the ore is blasted to be loose and then leaching liquid is injected to underground. The leaching liquid is collected and pumped on the ground and then metal is collected from leaching liquid. It is a cross-disciplinary technology integrating mining, ore dressing, and metallurgy.

## 3.3 Process Greening and Intensification of Oil and Gas Processing

### 1. Demands and Challenges

The processing of oil and gas resources is an important portion of chemical industry and plays an important role in the development of the national economy. The petrochemical industry accounts for 20% of the total industrial economy in our country and it is one of pillar industrial sectors. It is predicted that the demands of crude oil in China will come to 450 ~ 500 million tons by 2020, while the demands will reach 800 million tons by 2050. China is not only a big country of oil consumption, but also a big country in term of import dependency of oil and gas resources. At present, the rate of oil import dependency has increased to 55%. On the other hand, processing technical level of oil and gas resources in China is relatively low, suggested by high material and energy consumption, low utilization rate of resources, as well as serious pollution, greatly constraining the sustainable development of this industry. In addition, due to the problem of environmental pollution, the industries are unable to afford the huge investment needed during pollution end treatment, thought less effective for pollution control. Therefore, the application of basic principles of green chemistry and intensification technologies in modern process engineering is quite necessary. With original and integrated innovation, green process of oil and gas resource utilization should be achieved with respect to green material, cleaner process, and green product. In the next 30~50 years, the following areas need to achieve innovation and breakthroughs:

(1) Innovation in new mediums/materials in oil and gas resource processing.

(2) Enhancement in the intensification and system optimization in the refinery unit operation and whole process, improving the overall level and technology of current refinery.

(3) Development of green petrochemical industrial process, constructing green and clean modern petrochemical industrial system.

(4) Improvement of processing technical level and capability of heavy oil processing, achieving clean and high efficient resource utilization of crude oil.

(5) Emphasis of the integration and complementation of machining process for renewable energy/new energy and traditional oil/gas resources.

## 2. Objectives and Tasks

In terms of generic problems, significant problems, and cutting-edge problems in processing of oil and gas resources, fundamental and applied research should be carried out to improve knowledge innovative capability and engineering-based integrating capability. On the basis of the multi-scale structure involving materials, unit/processes, and system, molecular designing and manufacturing, process intensification, and reactor structure optimization designing for functional materials should be carried out; ubiquitous multi-scale theories and discrete methods should be developed; and the capabilities of material, equipment, and process quantify designing should be promoted. Based on industrial ecological principles, integration theory and methodology research for material-energy optimization should be proposed, green methods and lifecycle assessment methods should be improved, and integrated technologies of green process with high resources utilization efficiency, optimal economic benefits, and minimal influences on environment should be established.

The overall developing objectives:

Fundamental research, applied research, and technical developing level for oil and gas resources in China should be in the rank of advanced levels internationally, and China should have international competitive advantages in a number of areas.

(1) Develop a number of innovative world class oil/gas resource processing technologies and establish our oil/gas resource scientific R&D platform, setting up foundations for the sustainable development of our oil/gas

 *Advanced Manufacturing Technology in China: A Roadmap to 2050*

resource processing industries.

(2) Scientific and technical innovative capability, especially related to original innovation in quantify design and scientific and technical development for equipment and process in China should be improved dramatically. A number of scientific and technical achievements with independent intellectual properties are expected to achieve industrialization. The market competitive and industry leading technologies should be formed. The total investment should be reduced by more than 10% and the energy and material consumption will be reduced by more than 10%~15%.

(3) A group of world class talents with innovative scientific R&D capability should be trained to provide strong supports for the upgrading and transforming of the oil and gas resources treatment process as well as the developing of cutting edge and innovative technologies.

## 3. Key Technologies

### 1) Innovation of new medium/new material in oil and gas resources processing

The design, development, and application of new medium/new material to reduce reaction temperature, increase reaction efficiency, and decrease energy consumption of separation should be addressed. Especially, the research of novel catalytic materials and unconventional reaction medium should be strengthened and advanced and highly selective catalysts for the activation of C–C bond and C–H bond should be developed to obtain direct oxidation with minimal pollution and optimal atom economy reaction with nearly 100% selectivity. The fundamental chemical theories and methodology researches in the fields of chemical synthesis, catalysis and surface science, structural and functional materials, and supermolecular chemistry should be strengthened; the micro mechanisms of catalysis and separation process should be further understood; the design and synthesis of novel reaction medium and catalytic materials should be optimized; and the new generation technologies for oil and gas resources processing should be formed, enhancing the atom economy and resource utilization rate.

### 2) Unit intensification and system optimization of oil refining industry

The old equipments in existing oil refining industry have the disadvantages of low in reaction and transmission efficiency and high in energy consumption. Traditional oil refining equipments mainly use fixed bed, fluidized bed, and tubular reactor. The intensification of unit process is beneficial to greatly increase reaction and separation efficiency. The development and application of micro structure reactor with nano-mirco structure should be emphasized, and it is proposed that nano-based catalyst materials can fill in micro structure reactors to achieve accurate control for reaction temperature and reaction materials, achieving the objectives of high yield, selectivity, safety, and product quality. Separation process coupling reactor, membrane reactor, and external field (magnetic field, electric field, microwave, and etc.) intensive reactor should be addressed, integrating with computer simulated technology, to achieve reaction control and optimization at the molecular level. The whole system integration and optimization in oil and gas resources processing should be enhanced. Especially, integrated innovation of energy-material optimization based on industrial ecological principles should be intensified.

### 3) Upgrade of oil and gas resources cleaner processing

The production of clean oil, meeting the requirements of new environmental protection regulation and legislation, is still the focus of cleaner oil and gas resources production process. Its key technologies are to develop new generation of catalytic cracking and hydrogenation technology. The improvement of catalysts and technologies can produce low sulfur and ultra low sulfur gas, diesel oil, aviation kerosene, and the economic performance and reliability of diesel oil production with ultra low sulfur gas can be improved. Deep purification technology for new generation of non-hydrogenation oil and fiber membrane separation method should be developed to deal with the problems of material densification aggravation and hydrogen source shortage in future. Value-added and resource-based utilization of byproduct of light hydrocarbon in oil and gas resources processing should be enhanced and the new technologies of producing high value chemical products with light hydrocarbon material should be developed.

The process and value-added utilization of inferior crude oil should be emphasized. The technical level and capability of heavy crude oil processing should be promoted. The value-added and resource utilization of tertiary oil recovery and high sulfur heavy oil or oil shale should be emphasized. Processing technology for domestic oil, especially for the value-added utilization for of sands and oil shale should be developed. The new generation of catalytic hydrogenation cracking and catalytic refining technology should be developed to achieve cascade and value-added utilization of low quality resources. Gasification technology of heavy oil/residue or co-gasification technology with coal/biomass should be developed; the development of clean gasification of in-situ carbon sequestration should be addressed; and the objective of reducing discharge of greenhouse gas can be achieved.

### 4) The emphasis on the process integration of renewable energy/new energy and traditional oil and gas resources

The material and energy in traditional oil refining industry are almost all from chemical and petroleum resources, which have the disadvantages of high energy consumption, heavy pollution, and serious ecological damage. The utilization of renewable energy and new energy can solve the above problems. In particular, it provides basic way to the problem of energy resource shortage. Thus the ratio of the renewable energy and new energy in the oil and gas resource processing should be increased gradually. The utilization of photolysis water (waste water), hydrogenation, and biomass energy, as well as solar collector response system and other new technologies should be developed. The design, development, and researches of scale-up law for solar thermal/ photocatalytic reactor should be enhanced. On the one hand, the new technologies need to be integrated with traditional oil and gas treatment processes to provide adequate heat, electricity, and hydrogen source for oil and gas resources processing. On the other hand, according to resources features of China, non-petroleum-based chemicals (such as methanol, ethylene, propylene, and etc.) and alternative fuels (such as ethanol, biodiesel oil, and etc.) which will account for certain market share should be produced.

## 4. Developing

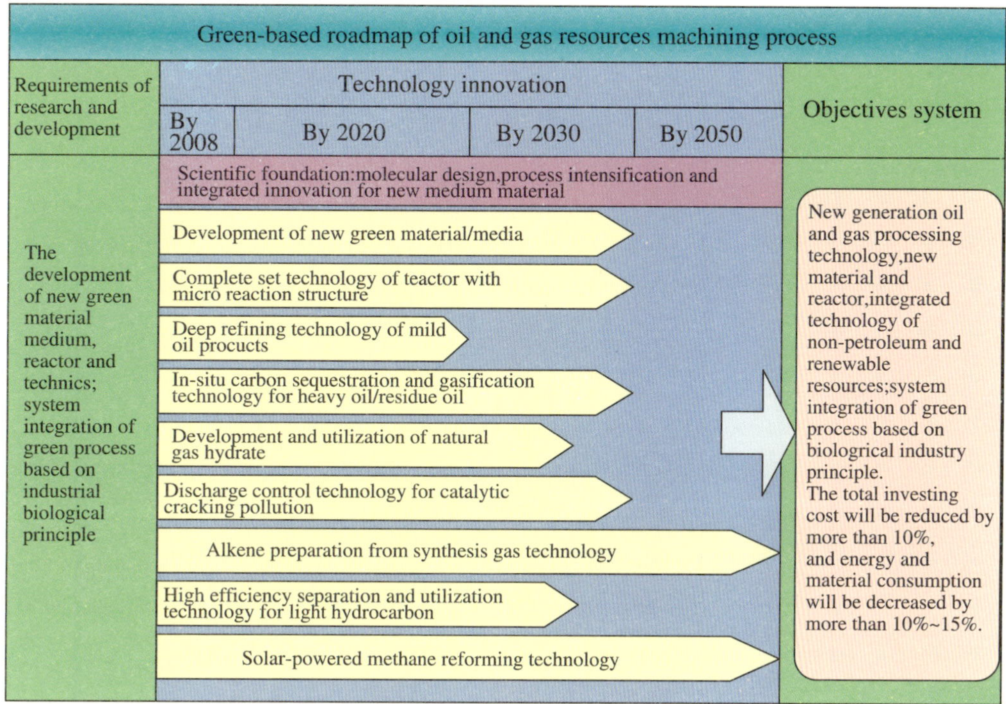


## 5. Summary

Cleaner oil and gas resources processing and process intensification is the requirements and development trends in the future. In the next 30 ~ 50 years, the new media/new materials in oil and gas resources processing, intensification and integration of oil refinery processing equipment, high efficient utilization of heavy oil, integration of renewable, new energy with oil and gas resources processing, as well as life cycle–based system optimization and industrial ecological design, and etc.,will achieve innovation and breakthrough. The conversion and utilization efficiency will reach optimal and its damage to resource and environment will be minimized at the same time.

In a word, the execution of scientific and technical developing roadmap will produce a series of original and integrated innovation technologies, such as reactor technology with micro reaction structure, deep refining technology of oil product, cracking technology for heavy oil, in-situ carbon sequestration and gasification technology for heavy oil/residue oil, natural gas hydrate and alkene preparation from methanol technology, alkene preparation from synthesis gas technology, high efficient separation and value-added utilization technology for light hydrocarbon, pollutants discharge control technology in oil and gas processing, solar collector reactor system, and etc. The development of these new technologies and the establishment of demonstration projects will greatly increase innovation capability of process science and engineering in China, especially for quantify design capacity and scientific and technical original innovation capacity for equipment and process. The industry leading technologies having market competitiveness will be formed to provide strong support for upgrading and development of advanced technologies in oil and gas resources processing.

> Green degree: it is a synthetic index quantitatively assessing environmental influences. It can quantitatively express the influences degree of material, logistics, energy flow, unit/process, and system on environment.
>
> The quantitative formula of green degree synthetically considers nine kinds of impact factors, including global warming potential (GWP), ozone depleting potential (ODP), acidification potential (AP), eutrophication potential (EP), ecotoxicity potential to water (EPW), ecotoxicity potential to air (EPA), human carcinogenic toxicity potential to water (HCPW), human non-carcinogenic toxicity potential to water (HNCPW). With the continuous emerging of new materials and development of environmental chemistry, environmental impact information included in green degree formula will be continually expanded and systematized.
>
> Green degree can be applied in environmental influence comparison of solvent screening of chemical process and different technological routes to provide information for design of chemical process. Meanwhile, as objective function of environmental impact, green degree will provide theoretical basis for multi-objective optimization of chemical process and become necessary and important tool of development of chemical process.

## 3.4 Secondary Recycling Utilization and Core Technology

### 1. Demands and Challenges

Traditional resource processing technology can only utilize 20% ~ 30% components of raw materials, while the rest has been released into the environment in forms of wastes, causing tremendous wasting of resources and environmental pollution, and severely affecting social sustainable development. For example, in China, water, coal, and iron ores are all important resources to the nation's economy, but the wasting recovering rates of these resources are relatively low. The recovery rate of coal resources is about 40%, and according to this rate, the wasting of coal resources will reach 56 billion tons by 2020. The reusing rate of industrial water in China is only about 30%, which is much lower than that in the developed countries (which is 75%). The overall recovery rates of mineral resources are relatively low, too. Every year about 11.5 billion tons of tailings containing average 11% of iron are discharged by metallurgical industries, equivalent to 16 million tons of residual iron in tailings. Furthermore, there are almost over 30 tons of residual gold in the over 20 million tons of gold tailings.

In addition, the amount of waste water, gas, and residue discharged every year is enormous. According to the national statistical bulletin of environment in 2007, the industrial wastes in China are 1.76 billion tons. The total amount of waste water is 55.7 billion tons, and industrial waste water is 24.7 billion tons. The emission of sulfur dioxide is 24.681 million tons. All of them are recoverable secondary resources, and their reutilization will be great beneficial.

Improving the secondary resource exploitation and utilization is also an important part of the solutions to release the pressure of resources shortage and environmental pollution in our nation. In 2003, the production of recycled aluminum is 8.08 million tons, which is only 27% of the production of refined aluminum. In comparison, the production of recycled aluminum in the United States is 2.93 million tons, which is 108% of refined aluminum, and the recycled aluminum in Japan is 194 times of refined aluminum. For the secondary lead, the United State's recycled lead accounts for 58% of its total refined lead production in 2003, and this number reaches amazingly 98% in France. The consumption of zinc oxide, zinc ingots, zinc powder, and zinc dust in western countries is about 6.5 million tons, among which 2 million tons are from secondary zinc resources. The secondary resource utilization rate in China is only 1/3~1/2 of the world's average, exhibiting great improvement potential.

The secondary resources recovered from wasted slag, dust, and tailings generated from process industries, and solid wastes produced in consumption process, will become the main source of resources in the near future. However, the technologies for secondary resources utilization are very complex. For example, the low grade of valuables in waste slag and tailing, the complexes composition in wasted products such as electronic waste, and the combination of organic and inorganic materials in many wastes, all attribute to the low separation efficiency of the secondary resources. At present, the recovery rates of these resources are very low and the processing technologies relies heavily on strong acid and alkali dissolution treatments, inevitably producing substantial secondary pollutions such as TCDD, heavy metal, ammonia, and acidic waste water. After years of dedicated research, many breakthroughs in the field of the secondary resource recycling and cleaning technologies have been made, including the biological metallurgy, vacuum metallurgy, and a variety of metal recycling technologies, etc. Next step, low pollution and energy efficient new secondary resources recycling technologies and equipments, as well as integrated technologies will become the research focus in the field of secondary resource recycling.

## 2. Objectives and Tasks

The goal of secondary resources utilization is to use resources from different sources and in various forms, manufacture new products with low energy consumption, save resources, prevent pollution, make practical contribution to the circular economy, and approach and realize the goal of establishing resource-saving society in China. The immediate objectives are: achieving primary comprehensive utilization of secondary resources; increasing the recycle ratio of metal to over 80%; establishing a complete technology system for the usage of exhaust, organic/inorganic compounds, and biological scraps; and forming industrial chain for secondary resources recycling.

In the next 20~50 years, there will be three expecting tasks for clean and efficient utilization of secondary resources. Firstly, to develop efficient and clean technologies for the utilization of residues slag and tailings produced during mining and ore dressing, as well as metallurgy slag and effluent during metallurgical processes, achieving shortened green separation of various metals and getting high-value products. Secondly, to develop recycling technologies for non-ferrous metal, waste mechanical & electrical products, and E-wastes, achieving efficient separation and recycling of multi-component materials containing metal, organic, and inorganic substances. Thirdly, to develop technologies for the advanced application of the multi component resources and for the cost-effective enrichment of high-value components with low concentration, achieving high-value utilization of biological residues and waste gases. It is expected that in the next 20~50 years breakthroughs will be made in key technologies for clean and efficient utilization and recycling of secondary resources.

## 3. Key Technologies

The recycling technology of secondary resources is a comprehensive technology integrating separation science and technology, catalytic reaction technology, biological technology, and process integration technology. The research targets are industrial and household wastes in the form of gas, liquid, or solid. The research purpose is to separate the useful components in the materials to make them into high-grade resources or transform them into new products under certain conditions.

The environmental technology referred in this section emphasizes avoiding pollutants production or reducing the total amount of pollutant production  and their environmental toxicity by producing decomposable pollutants with the utilization of  environmental self-purification capability in industrial production process and secondary resources recycling process.

Due to the complexity of secondary resources and environment problems, the comprehensive recycling and utilizing of all the wastes such as the complete separation and analysis of the whole components, the trace component recycling, the low cost detoxification, and etc., still have many challenges; however, breakthroughs are expected to be made in the key technologies as follows.

### 1）Multi-metal shortened green separation and recovery technology

Scrap metal, alloy, tailings, metallurgical chemical residues, sludge, dust, and other secondary resources generally contain various metals, and hence multi-metal short-range green separation will be the focus of technology research in the next 20~30 years. The intelligent automatic sorting, efficient scrap metals refining, new alloy selective leaching, efficient new extractants, and microbial selective fixation techniques will make breakthroughs. For example, vanadium and chromium clean separation technology of industrial heavy metal slag; advanced separation technology of vanadium, titanium, and iron in vanadium steel slag and high titanium-bearing slag; and highly efficient separation and waste recovery technology and equipments for toxic heavy metals, noble metals, and rare earth metals in multi-component industrial solid wastes will achieve technical breakthroughs and industrial application. The whole flow integration and optimization, the zero residue discharge of the tailing, and waste recovery process will be achieved. In the next 30~50 years, the metal recycling utilization rate will rise from 20%~25% to 80% in China.

### 2）Sulfur recovery technology in metallurgical industry

Sulfur containing gas and liquid wastes are main pollutants in metallurgical processes, and the fundamental approach to eliminate their environmental effects is sulfur recycling. The current production of sulfur in the world mainly comes from the natural sulfur, acid gas, sour crude oil, pyrite, and non-ferrous metal sulfides; while the amount of the recovery sulfur will become increasingly larger proportion of sulfur resources in the future. In the next 20~30 years, efficient desulfurization catalysts and new bio-desulfurization technologies will play important roles, and in the next 30~50 years more than 85% of the sulfur supply in the world will come from the recovery sulfur.

### 3）The recycling technology of waste gas

In the next 20~50 years, significant breakthroughs will be made in the field of green separation and chemical cleaning technology for utilization of industrial waste gases including $SO_x$, $CO_2$, or $C_xH_y$. Due to the successful application of efficient separation medium (such as extraction membrane materials and  physicochemical absorption medium), highly efficient chemical treatment, and biological catalysts, using waste gas to manufacture sulphur and sulphuric acid directly is expected to become a new industry, eventually realizing zero waste gas discharge.

### 4） The recycling technology of waste water

In the next 20~40 years, the technologies of using low concentration metal ions in waste water to prepare high-value products directly, new membrane materials for membrane distillation, equipments for membrane distillation, and highly selective novel ion exchange resin and green extractants are most expected to achieve breakthroughs. By recycling water resource comprehensively and utilizing solute to obtain high value products, waste water zero discharge can be realized.

### 5） The technology of high value utilization of biomass residue

In general, besides sugar and trace amount of inorganic components, biomass residue mainly consists of lignin, protein, and fiber. Green technologies such as steam explosion, high catalytic cracking, enzyme hydrolysis, and novel ionic liquid medium for handling biological residue to prepare biodiesel, methanol, ethanol, glucose, protein, and cellulose derivatives are expected to obtain a breakthrough and get wide industrial application in the next 20~30 years.

### 6) Waste recovery of waste organic -inorganic composite

Organic-inorganic composite waste is a new type of waste originated from modern industries, existing widely in the PCBs, medicines, fertilizers, catalysts, membrane materials, and other new materials. In the next 20 years, the low-cost green separation technology for the separation of organic and inorganic components from composites is expected to achieve a breakthrough, and in the next 30 years, high-value and the comprehensive utilization of waste organic-inorganic composites, PCBs as representing, technology is expected to realize industrialization.

### 7) Optimized integration of whole tailing and slag treatment process

Process optimization is capable of balancing materials, water, and energy precisely, and energy conservation and consumption reduction can be realized by optimizing and integrating the whole tailings and slag utilization process. The integration and optimization of the complete process technologies covering tailings sludge collection, pretreatment, separation, purification, and recycling technology will be fully completed around 2050 based on the optimization for each unit operation.

### 8) Industry chain of secondary resources recovering

Utilization of secondary resources alone is often expensive and may generate new wastes, becoming secondary pollution. Through comprehensive layout and ecological planning, it is expected to establish the industry chain for secondary resources recovering and achieve flexible convergence of upstream and downstream. Industry chain of secondary resources recovering and the ecological industrial park for utilizing of all components are expected to be established by 2050.

  *Advanced Manufacturing Technology in China: A Roadmap to 2050*

# 4. Developing

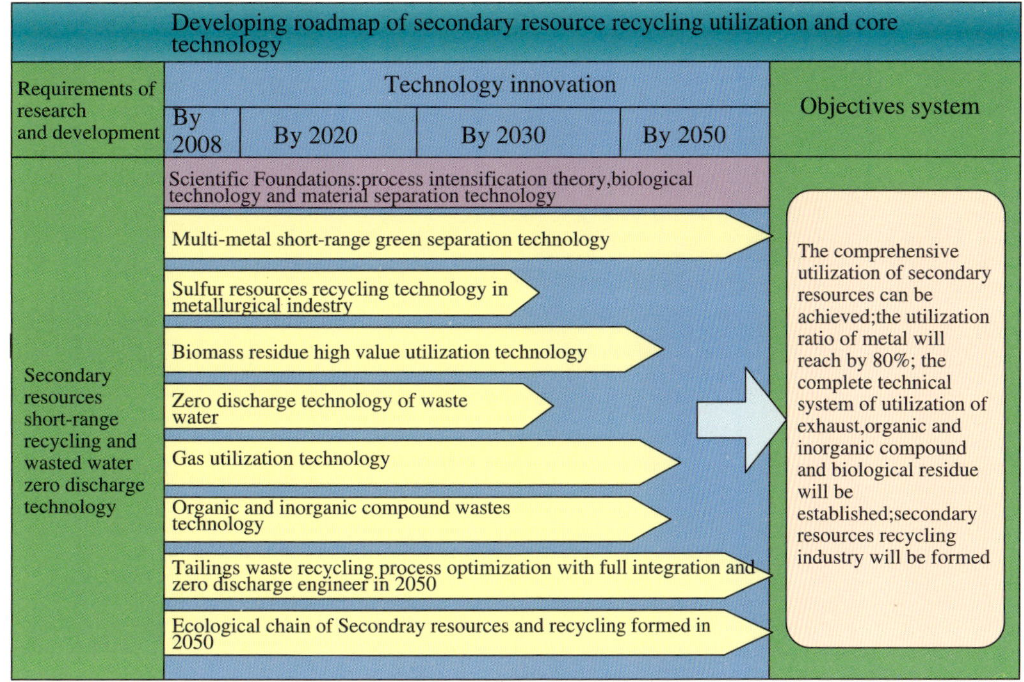



# 5. Summary

In summary, sulfur resource recycling technology for metallurgical industries, especially the technology for high value utilization of sulfur in smoke, as well as the technology for comprehensive utilization of biological residual and waste water will be established by 2020. The technologies for advanced application of complex components in biological residual, waste water/gas, and organic/inorganic composites; for efficient recovery of the low concentration precious components; and for short-range green separation of selected metals will be formed by 2030. By 2050, the technologies for short-range green separation of metals will be fully developed; the technologies for efficient separation of multi-component metallic, inorganic, and organic materials, as well as all components recycling will be widely applied; zero emission tailing and slag utilization projects with whole process integration and optimization will be built; and ecological chain of secondary resources recycling will be established, achieving more reasonable regional industrial structure.

## 3.5  Advanced Process Simulation, Integration, and Optimization Technology

### 1.  Demands and Challenges

After the 21st century, industrial production is facing challenges from economic, environmental, and social aspects. Globalization greatly promotes the free mobilization of human, technology, capital, information, and products beyond limitation of countries. It brings market opportunities and it also calls for advanced production technology with globally competitiveness. Besides higher performance and higher qualities products, living environment also need to be improved. This demanding increases the requirements for energy saving, material saving, and environmental friendly green production technology. In addition to the traditional experimental research methods, green manufacturing industrial fields are increasingly using advanced process simulation, integration, and optimization technologies.

Simulation technology has achieved rapid development in past 30 years. The sustained growing industrial demand has been its driving force. For instance, CFD is used in regress design, analysis, and optimization in the field of aviation, aerospace, automotive field, and etc., becoming standard steps and methods. The application of computing technology have been continuously broadened to many fields, and further energy, chemical industry, metallurgy, electronic industry, food processing, environmental protection, municipal construction engineering, construction, mechanical and electrical engineering, medicine, and etc. gradually use computing as basic designing and analysis means.

With rapid development of computer technology, computing speed of

high performance computer is increasing exponentially, establishing foundation for wide application of process simulation technology in manufacturing industry. The research of process simulation technology seems relatively lagged, and is far behind developed countries including Europe, the United States, and Japan now, as exemplified by the fact that main computational fluid dynamics commercial software, e.g. Fluent and CFX, all belongs to American and British companies. Compared globally, the investment in simulation technology in China is relatively low according to our GDP.

## 2. Objectives and Tasks

Numerical simulation technology combines computational mathematics, computer science, fluid dynamics, scientific visualization, and many other disciplines. It uses modern computer technology to achieve fully digitizing. The accurately simulated results are obtained to analyze and optimize the existing process and provide basis for further process scaling-up. Compared to the traditional physical test, simulation technology has obvious advantages and is being applied in more and more disciplines. Its application prospect is extremely broad, e.g. chemical reaction flow, multiphase flow and transport, high temperature radiation, and soft matters.

For the process of manufacturing industry, especially for the increasing complexity of process manufacturing, process simulation, integration, and optimization technology need to solve computing precision and computing efficiency problems in large-scale complex system, as well as extreme condition computing problems. The overall objectives of advanced process simulation, integration, and optimization technology are to shorten producing cycle and optimize manufacturing techniques in order to increase production and save energy and resources, and further, to reduce the wastes and their  toxicity; to design new product and new production techniques effectively to achieve environmentally friendly safe manufacturing and improve ,  human health simultaneously. Immediate specific objectives are: shorten the time from laboratory research to industrial applications by 30%~50%; decrease the investment of equipment by more than 10%; and increase separation efficiency and decrease materials consumption by 10%.

## 3. Key Technologies

Process simulation, integration，and optimization technologies which are important methods to achieve green manufacturing industry，are 'virtual' manufacturing process experiments carried out on computers, and will take full consideration of the environmental impacts on the optimizing design in the future. Compared to the traditional experimental and empirical methods, the simulation technologies are mainly superior in the following aspects:

(1) Based on experimental data, simulation technologies can repeat the same experiment, or simply modify experimental parameters to obtain more test results.

(2) Due to the complete control of the test objects and environments, the time needed to synchronize the physical conditions including test objects and environments is greatly reduced.

(3) Simulation technologies only need direct operation by designers such as engineers, reducing the manpower cost of physical manufacturing in the early stage of experiments, thus, significantly saves human resources.

(4) Due to the complete control of the experiments with computer, the potential physical harm to human and pollution to environments are avoided.

Although there are challenges such as cross-disciplinary of multiphase fluid mechanics, computer capability, physical and mathematical modeling of engineering problems, and etc. in simulation technologies, it is promising to have breakthrough in the fowling aspects:

### 1) Parallel computing, parallel algorithm, and parallel programming

Develop 'populace-oriented' high performance parallel cluster systems, which are suitable for manufacturing engineering simulation by 2020. Develop high efficiency, high precision, large-scale, transplantable parallel algorithms, and parallel programs, accomplishing the research and development of multiphase flow, transmit, and reaction process related advanced computing technologies for chemical engineering, petrochemical engineering, metallurgical engineering, and material sciences by 2020.

### 2) Molecular simulation technology

Molecular simulation technology integrates quantum mechanics, Monte Carlo methods, molecular mechanics, molecular dynamics, and etc.; uses atomic level molecular model to simulate molecular structure and behavior so that various physical and chemical properties of molecular system can be achieved. Molecular simulation can simulate static structure and dynamic behavior of molecular, as well as physical phenomena and processes which still cannot be examined by any modern physical experimental methods. It can also study chemical reaction path, transition state, reaction mechanism, and other key problems. Further, it can simulate a variety of spectrum in molecular system, and carry out products structural analysis and designing. Since the 1990s, molecular simulation technology has widely applied in the molecular sieve catalyst, polymer materials, inorganic materials felids, and etc. For instance, Mobil and other companies are positively applying molecular simulation technology to impel the R&D of new products and new materials. At the beginning of 1990s, molecular simulation technology entered a new stage in the application in material and product design fields. It does not only provide a qualitative description, but also gives the quantitative results for some structure-performance relation of molecular systems. The development of molecular force field, simulation algorithms, and computer hardware establish solid foundation for the development of molecular simulation methods. It is expected that during 2020~2030, molecular simulation technology will be widely applied in green chemical field.

**3) Turbulent flow modeling and high precision numerical computing**

Multiphase turbulent flow is a complex object that the process industry has to face during the design of processing equipments. Existing numerical simulation methods such as direct numerical simulation (DNS), large eddy simulation (LES), and Reynolds-averaged simulation (RANS) have their advantages and disadvantages. Irregular boundary treatment technology is one of the keys. The research of body-fitted coordinates and non-structured grid technology will be enhanced. Around 2020, the adaptability of irregular geometry boundary should be developed and irregular mesh generation technology which is suitable for parallelization with short grid time should be generated. According to engineering application demands, highly efficient turbulence model and simulation computing methods for analysis, scale-up, design, and optimization should be founded before 2030.

**4) Mechanism model and simulation technologies for multiphase flow, transport, and reaction process coupling**

For ubiquitous multiphase reaction system in chemical engineering, petrochemical engineering, metallurgical engineering, and material science fields, as well as flowing, transmission, and reaction process in multiphase reaction and separation equipments, multi-scale model and simulation research should be carried out; the physical and mathematical models in multiphase coupling process based on mechanism should be established; and the highly efficient computing method of engineering simulation should be developed. Scientific simulation computing of the large mixing tank, fluidized bed, fixed bed, multi-phase bubble column, and other industrial equipments, as well as engineering application in reaction separation processes should be achieved by 2030.

**5) Green process system integration technology**

Originated in 1970s~1980s, process system integration technology has been developed to minimize energy consumption by optimizing and integrating technical processes as a whole. Green process system integration technology is the integration of information technology and process industry, and its developing direction is multi-scale green process system integration. Taking materials-energy- information flow of the process system as research objectives, green process system integration technology studies how this kind of system meets the demands for optimal economy development and environment protection simultaneously by resources allocation, planning, management, design, and operation control. The objective is to achieve the optimization of sustainable production on technology and structure. The range of research has covered many levels such as molecular design, process optimization, and ecological industry park system integration. The key developing technologies include: molecular design and product engineering; micro-scale optimization and integration; process intensification and intelligence-based equipment;, process optimization and integration for large complex system; bio-industry

system integration in macro-scale; green supply chain designing and optimization; and etc..

### 6) Advanced engineering analysis, scale-up, design, system integration, and optimization tools

The integration of computer aided software would be enhanced with the maturity of simulation technology. The full lifecycle simulation and optimized integration platforms for discrete manufacturing and process manufacturing which covers the resource extraction, manufacturing process, product design and consumption, and waste recycling processes should be formed by 2050. Process simulation technology will replace part of pilot tests, and provides scientific basis for quantitative amplification of new technology process.

### 7) Business simulation software and the automatic function

Develop large-scale homemade numerical simulation commercial computing software by 2050. The software could automatically achieve grid generation, grid adaption, assessment of sensitive parameter inputting, optimized designing, and continuously updating expert bank based on large number of calculation practice-known as knowledge engine technology and multi-field coupling. It is expected that multi-scale coupling simulation computing from molecule, particle, equipment, and process to microscopic-mesoscopic-macroscopic industrial park will be realized.

## 4. Developing

| Requirements of research and development | Technology innovation | | | Objectives system |
|---|---|---|---|---|
| | By 2020 | By 2030 | By 2050 | |
| Green-based, energy saving and consumption reduction of resources conversion utilization, design, engineering amplification and control of new equipment and new technics | Scientific foundation:process intensification theory,multi-scale model and computing method in green manufacturing | | | The time of from laboratory to industrial applications will be shortened by 30%~50%; factory investment in equipment will be decreased by more than 10%;separation efficiency is increased by more than 10% and energy and materials comsumption will be reduced by more than 10% |
| | Parallel computer,parallel algorithm and high performance computing | | | |
| | Molecular simulation computing technology | | | |
| | Turbulence model and high precision numerical computing technology | | | |
| | Multiphase complex reaction process and equipment model and simulation technology | | | |
| | Green process system integration technology | | | |
| | Engineering analysis,amplification,design,system integration and optimization tool | | | |
| | Commercial simulation computing software and automatic function | | | |

## 5. Summary

In summary, high precision large-scale parallel computing process simulation technology, especially for computational fluid dynamics simulation technology of complex process system, will be formed by 2020. Process simulation technology of alternative resources processes under extreme

conditions should be established by 2030. The full lifecycle simulation and integrated optimization platform covering discrete manufacturing and process manufacturing which include resource extraction, process manufacturing, product design and consumption, and waste recycling should be formed by 2050. Process simulation technology will replace parts of pilot tests, which provides scientific basis for quantitative design and scale-up of new techniques process. Finally, advanced process simulation, integration, and optimization technology will be widely applied in green manufacturing industry process.

CFD Technology: CFD is short of Computational Fluid Dynamics. It develops with the development of computer technology, numerical computing technology. Simply, CFD is 'virtual' computer experiment, which simulates the actual fluid flow, heat transfer, mass transfer and other related phenomena. A complete CFD simulation includes the following steps: firstly, establish mathematical model reflecting problems (engineering problems, physical problems, and etc.) and set up differential equations and corresponding fixed solution conditions describing relationships among variables in the problems; secondly, develop highly efficient and highly precise computing methods, e.g. discrete mathematical equations and solution methods and establishment of computing grid and process of boundary conditions; thirdly, prepare program for simulation computing and output the results (such as the calculated flow field), improving the equipments and techniques and optimizing design, analysis,  and control according to the simulated results to achieve high efficiency, green, and clean production. For simulation of complex engineering problems, experimental validation is necessary. With the development of computer image display software, image display of numerical results will be rapidly developed to be fast and timely, capable of 3D scanning and image reality, and etc. Dynamic process is displayed with playback equipment and numerical simulation can give full play to numerical experiments.

## 3.6  Biomass Processing and Bio-engineering Technology

### 1. Demands and Challenges

Biomass refers to all the organic substances converted from the solar energy and carbon dioxide originated from photosynthesis. Also, the animal substances and their metabolites such as agricultural products and crops, straw, wood and wood waste, animal waste, urban wastes, aquatic plant, and etc., which are created by photosynthetic products indirectly can be included as biomass. Biomass resource is the natural solar energy carrier in which energy stored in chemical form, and it is the most abundant renewable resource on earth. Every year, it may produces $(1.6\sim2)\times10^{11}$ tons of biomass that contain about $3\times10^{21}$ J energy, equivalent to 60 billion to 80 billion tons of oil, or in another words, 20~27 times of global oil consumption each year. The effective development of biomass resources can completely satisfy the human needs for energy and resources.

At the beginning of the 20th century, the majority of the non-fuel industrial products, including dye, paints, medicine, chemicals, cloth, and fiber were basically obtained from biomass. But in the 1970s, the organic compounds derived from oil replaced the biomass products substantially, possessing more than 95% of the market. In terms of the energy consumption ratio, oil consumption account for more than 70%. After years of mining, oil resources are gradually depleted. In addition, the extensive use of fossil fuels caused prominent problem of greenhouse effect and other environmental problems, deteriorating the human living environment. Recognition has been achieved that in order to establish a sustainable development of human society, renewable biomass resources must be used to replace oil resources to meet the needs of energy, materials, and chemicals. In this context, scientists and engineers are working to develop new technologies to reduce production cost of bio-based products and increase the application scope of bio-based products. At the same time, governments and the people around the world also paid more attention to the development and utilization of bio-based resources. New bio-based resources are continuously promoted via enterprises and research institutes through the environmental and legislation measures.

In recent years, a variety of biomass conversion technologies for the development of the biomass resources utilization has made progresses. However, the use of biomass is still in the research and development stage in general. This technology has not made any fundamental breakthroughs, and is far from large-scale industrial production. In this regard, there are great challenges mainly from the following aspects: firstly, it is difficult to use biomass completely due to the complexity of the biomass material composition, thus difficult to separate. Usually, only single lignocellulose composition such as furfural, paper, and xylitol from lignocellulose can be used, and other compositions will be discarded as wastes; not only resources wasting, but also environmental polluting. The biomass material did not improve the economic efficiency, but burdened it. Secondly, biomass processing involves multidisciplinary integrated with the biological, chemical engineering, chemistry, and engineering, due to the fact that single technology or single discipline will not be much help. It is necessary to break the subjects' boundary and establish concept for cascade utilization of material resources without emphasizing one-sided thinking on biological method or heat chemical method. Thirdly, the large scale utilization of biomass resources is not only limited due to their regional and seasonal distribution, but also suffers from the contradiction of resource decentralization and industrial centralization production. Therefore, the utilizing of bio-based

    *Advanced Manufacturing Technology in China: A Roadmap to 2050*

products should be fully coordinated with agriculture and forestry.

## 2. Objectives and Tasks

Biomass processing faces two major tasks: the first one is to develop substitutions of fossil fuels and chemicals, such as fuel ethanol, bio-diesel, and other chemicals derived from fossil-based raw materials. The renewable resources need to be developed to satisfy people's needs because there will be a final end for the limitation of fossil fuels. The other one is to develop bio-based new functional products. In order to meet the social developing and the material demanding due to people's living quality improvement and diversity, new products that could satisfy the consumption needs should be exploited. In general, bio-based products include bio-fuels, bio-based chemicals, and biological materials, and 40% of the fossil fuels are expected to be replaced by bio-fuels and 45% share of the total market will be replaced by bio-based chemicals and biological materials by 2050. For this purpose, the following key technological research must be strengthened in the next 30~50 years: High-performance and low energy consumption pretreatment technology, high-performance low-cost lignocellulose producing technology, new media of ionic liquids technology, modern biotechnology, lignin modified processing technology, as well as the integration research of these technologies. Finally, an ecological industrial chain is expected to be established for the full utilization of biomass.

## 3. Key Technologies

### 1) High-performance & low energy consumption pretreatment technology

Pretreatment is the starting step of biomass processing, and it is a continuous work to develop high-performance and low energy consumption pretreatment techniques for different biomass and subsequent conversion technologies in the coming decades. The pretreatment technology of straw and other lignocellulosic has been studied extensively, and the acid and steam explosion pretreatment have been industrialized. Because of the severe requirements for the equipments and heavy pollution, acid pretreatment technology is gradually eliminated. Steam explosion pretreatment technology is an ideal technology due to less pollution and low energy consumption. However, it still needs to be improved in areas such as developing of continuous steam explosion technology, reducing energy consumption, and improving efficiency. Microwave technology and radiation technologies are also gradually applied to straw pretreatment. It is expected that straw pretreatment energy consumption can be reduced by more than a half by 2020.

### 2) High-performance & low-cost cellulase production technology

The high cost of cellulose is the major obstacle for the biological conversion of lignocellulose. Take cellulosic ethanol production as an example, it is reported that the minimum cost levels of cellulose hydrolysis can be controlled to be 10~20 cents/gallon of ethanol, while the starch-based ethanol costs only 1~2 cents/gallon. It is therefore important to reduce the cost of cellulase enzymes that turn lignocellulose materials into large microbial product. It is expected the cost of cellulase enzymes will be reduced by 4~5 times through screening and reconstruction of strains and production process innovation by 2020~2030.

### 3) New media of ionic liquids applied in cellulose for high-value product manufacturing technology

There is high energy consumption, processes complex, and serious environmental pollution problems in traditional cellulose processing technology. The key problems for cellulose chemical modification are developing new non-polluting green solvents. It will be great break though to traditional technique for cellulose-base material production by utilizing non-volatile and cellulose diffluent ionic liquid as the medium in producing functional regenerated cellulose fiber and functional cellulose derivatives. By 2020, it is expected that the ionic liquid cleaner technology prepared for cellulose materials producing can realize industrialization.

### 4) Modern biotechnology

Screening out the excellent strains is the key in bio-transformation. Through the screening and tame of strains, combining modern genetic engineering technology, metabolic engineering technology, and genomics technologies, strong resistibility strains which are adapted to produce target production from the lignocellulose will be constructed and screened out. It is expected to construct yeast that equally use five carbon sugar and six carbon sugar to produce ethanol by 2020.

### 5) Lignin modifying and processing technology

Lignin is a kind of cellulose aromatic polymers, and the quantity of which is just lower than nature cellulose. As the natural random copolymer phenols, it is constructed by benzene propane unit through the ether bond (about 2/3) and C–C bond connected together with a three-dimensional body structure. Every year, about 60 billion tons of lignin can be produce globally. With many different types of active functional groups, lignin molecule is believed to be a great green chemical material due to its superior properties, such as renewable, biodegradable, nontoxic, and low cost. Currently lignin processing technology have not achieved substantial progress, most lignin was used as fuel or simply discarded. In the next few decades lignin processing technology will be significantly improved, and breakthroughs are expected in integrating lignin with phenolic resin, polyurethane, polyolefin, rubber, polyester, polyether, starch, soy protein, and other organic complexes via chemical reaction and physical blending to produce special functional polymer materials.

### 6) Bio-refinery technologies for mass production of chemical through

The production of ethylene, propylene, and other basic chemical raw materials are heavily dependent on oil resources. With the shortage of oil resources, replacement of oil refining with bio-refinery is an inevitable demand for social development. By a series of similar petroleum-based technologies such as dehydration, polymerization, cracking, hydrogenation, and aromatization, commodity chemicals will be produced. According to the prediction, by 2050, 45% of the fossil-based chemicals will be replaced by bio-based products; by 2025, bio-ethylene can be achieved; and by 2035, bio-propylene can be mass production and marketing application.

### 7) Technology integration of biomass comprehensive utilization and the establishment of ecological industry chain

Differences of biomass materials, as well as diversity of conversion technologies and conversion products make the conversion process a complicated engineering system. It needs to develop new technologies in various fields and integrate various technologies effectively to form ecological industry chains; to produce economic, social, and ecological benefits. In fact, the development and utilization of biomass resources involves many disciplines and multiple industries, and their development and utilization does not mean a simple piecing together the fragments following some of the above disciplines, but establishing a multi-disciplinary network of sciences, technologies, and productions. The process of biomass conversion integration technology follows these principles:

(1) Using and exhausting of each component of the biomass materials based on multi-layer classification methodology. According to the characteristics of components of plant, design optimal technical route, allowing each component to be able to transform into the final product.

(2) The lowest energy principle. Based on the design or shaping of equipments, combine heat and energy utilization, energy cascade utilization, and other methods to maximize energy utilization efficiency.

(3) Developing original and innovative technology by integrating various old and technologies. Form the ecological chain, the final pattern for biomass resources development, and it is expected that by 2030~2050, preliminarily ecological chain of biomass resources will be established.

The development of these key technologies is expected to be target product-oriented, and in the next 30~50 years, the main target products are bio-ethanol, bio-butanol, bio-ethylene, bio-propylene, and bio-degradable plastics such as poly-hydroxy fatty acid ester ( PHA) and so on.

## 4. Developing Roadmap

| Developng roadmap of biomass processing key technology and product development | | | | | |
|---|---|---|---|---|---|
| **Requirements of research and development** | **Technology innovation** | | | | **Objectives system** |
| | **By 2008** | **By 2020** | **By 2030** | **By 2050** | |
| **Utilization of biomass resources into the green process engineering** | Science foundation:High conversion process engineering based on the characteristics of biomass materials | | | | Established eco-industry with the center of renewable biomass resources,reduce dependence on fossil resources gradually,and realize the sustainable development of human society |
| | High-performance& low-power pretreatment technology of biomass materials | | | | |
| | High-performance cellulase production technology | | | | |
| | Modern biotechnology | | | | |
| | Anaerobic fermentation technology | | | | |
| | New media of ionic liquids applied in cellulose high-value production product technology | | | | |
| | Enzymatic hydrolysis,fermentation,separation coupling technology | | | | |
| | Non-oil route bio-chemical refining technology of chemical production | | | | |
| | Technologies integration of biomass comprehensive utilization | | | | |
| **Products and Industrialization** | Sweet sorghum ethanol,cassava ethanol,cellulose ethanol industrialization | | | | |
| | A large-scale industrialization of bio-butanol in 2030 | | | | |
| | A large-scale industrialization of bio-ethylene in 2025,A large-scale industrialization of bio-propylene in 2035 | | | | |
| | A large-scale industrialization of PHA in 2030,replaced 20% of fossil-based plastics in 2050 | | | | |
| | By 2050,bio-fuels alternative 40% of fossil fuels.Bio-based chemicals and biological materials account for 45% share | | | | |
| | Formation of multi-product generation eco-industrial chain | | | | |

## 5. Summary

The environmental pollution problems caused by extreme using of limited fossil resources have become increasingly prominent. Consequently, governments have gradually paid more attention to the development and utilization of renewable biomass resources. A large number of funding and scientists from various disciplines are coordinated and integrated to work on biomass research, which is believed to be the key for the development of biomass processing technology in the next 30~50 years. It is critical for us to thoroughly understand the diversity and complexity of biomass, emphasize interdisciplinary and utilization of multi-technologies, produce diversification of bio-based products, and establish an ecological industry chain with renewable biomass as the core. So that the sustainable development of social economy will be realize.



## 3.7 Utilization Technology of Low Carbon Resources and $CO_2$ Reclamation

### 1. Demands and Challenges

As important resources, low carbon resources have been addressed significantly in the 21st century. The corresponding low carbon chemical industry mainly focus on chemical processes and techniques for preparing compounds with two or more carbon atoms using synthetic gas (converted by natural gas, coal bed methane, coal, heavy oil, and etc.), methane (coming from natural gas, coal bed methane, and etc.), $CO_2$, and coal as raw materials. Currently, as a type of non-renewable resource, petroleum is decreasing gradually, while natural gas and coal will play an important role in energy resource structure gradually. Therefore, low carbon chemical industry, which takes coal and natural gas as raw materials to produce liquid fuels and chemical products, is the developing direction for future energy resource and chemical industry. The development of low carbon resources may bring revolutionary changes for basic organic raw materials. Due to the shortage of petroleum resources in China, the development of petroleum substitutes and its derivative substitutes is significant in terms of economic development and strategic deployment. Coal resources are non-renewable fossil resources, and thus as the inevitable technical strategy, it is important to highly efficiently utilize and optimally exploit these resources, meeting the demands of oil, gas, and chemical products through non-petroleum route.

In addition, extensive consumption of petroleum, coal, natural gas, and other fossil energy resources in short period leads to rapid increasing of $CO_2$ discharge which brought in greenhouse effect. The catastrophic climate changes have given a warning to human. There will be serious influences if no effective means are carried out to reduce $CO_2$ discharging and isolate $CO_2$, and the concentration of greenhouse gases such as $CO_2$ will reach the maximum at the end of this century under current trend. Thus, the new technologies for discharge reduction and utilization of $CO_2$ are also becoming increasingly urgent. As proposed by Japan, $CO_2$ discharge will be reduced by 50% by 2050. The construction of low carbon society will be the common goal of human society for the next few decades, a century, or even longer.

## 2. Objectives and Tasks

During the 30 years of reforming and opening-up, Chinese economy has made remarkable achievements. However, the rapid increase of economy leads to the increasing demands for energy resources. Currently, China is facing the serious problems of energy resources shortage and the heavy import dependence, seriously affecting the national security. Meanwhile, as one of the responsible large countries, China will take effective measures to reduce $CO_2$ discharge. The increased utilization rate of low carbon resources is one of effective methods, and reduction of the $CO_2$ discharge is also imperative for sustainable development of economy. The immediate developing objectives on low carbon resource utilization and $CO_2$ utilization are as follows: based on the development of new technologies, it is expected to achieve highly efficient utilization of low carbon resources through coupling cascade utilization technology, and the utilization rate of low carbon resources will be increased by 20%; large-scale cost-effective $CO_2$ separation will be realized in pillar industries. With the promotion of advanced $CO_2$ resource utilization technology, the internal recycling and utilization of carbon will be initially realized in the field of low carbon resources.

To achieve aforementioned objectives, the short-term tasks mainly include:

(1) Breakthrough of key technologies involving highly efficient activation of C–C bond, C–H bond, and C–O bond under moderate conditions; the development of $CO_2$ absorption medium with high adsorption capacity and large-scale cost-effective production technology; and the system integration technology of $CO_2$ resources.

(2) System integration and large-scale application of mature technologies including ultra-large-scale methanol production technology and large-scale coal-based poly-generation technology.

## 3. Key Technologies

Scientific and technological development and the pressure from energy

   *Advanced Manufacturing Technology in China: A Roadmap to 2050*

and environment greatly impel the industrialization process of low carbon chemical industry. Due to years of accumulation, substantial achievements have been made in China. For example, the coal gasification and FT synthesis technology has gradual matured, as suggested by dramatic decrease of unit investment. Today the main objectives of low carbon chemical industry are to develop carbon cascade utilization technology of resources; synthetic gas to synthetic liquid fuel technology; synthesis gas-based chemical production of non-petroleum route; as well as cost-effective $CO_2$ capture, separation, sequestration (CCS), and green utilization technology (CCUS). In the next few decades, the key technologies of low carbon resources utilization and green $CO_2$ utilization mainly include the following aspects.

### 1) Cascade conversion technologies for high-value integrated utilization of coal

At present, combustion is the main way to consume coal, accounting for about 85%. Coal is the unique fossil resources containing a large number of aromatic hydrocarbon structures. Current combustion and other commercial gasification technologies would completely destroy coal structure and therefore cannot achieve full utilization of the chemical materials including basic aromatic hydrocarbon structure, e.g. benzene and methyl benzene, which are still from petroleum refining. But the shortage of petroleum resources urges the development of alternative production routes for basic aromatic hydrocarbon material. The cascade coal conversion and high-value coal utilization technologies featuring the utilization of aromatic hydrocarbon structure and energetic characteristics of coal to achieve multi-production of chemicals and electricity (or heat), as well as the utilization of the inherent structure and the C/H elements of coal to achieve multi-production of aromatic hydrocarbon chemicals and synthesis gas, reflect the strategic need of our nation in the area of energy resources. In this field, China has already established a solid foundation and with further efforts in the next 5~10 years, technologies and techniques for the aromatic hydrocarbon materials and thermal/electric poly-generation as well as the aromatic hydrocarbon and synthesis gas poly-generation will be formed and the industrialized consequently .

### 2) High efficiency catalytic conversion technology for C1 chemistry

Abundant synthesis gas materials can be obtained through coal gasification and reformation of natural gas and $CO_2$, and the key of high-value utilization of synthesis gas relies on the development of high efficiency catalyst technology. Based on the core catalyst researching, with the development of new materials, new equipment, and new technology, the C1 chemical industry will be featured as low energy consumption and highly efficient catalytic conversion, exemplified by the utilization of synthesis gas. This technology, targeting at national strategic demands, is expected to make breakthroughs in the area of high efficient catalyst for synthesis gas conversion, directly preparation of alkene and coal-based fuels (Including FT oil, low carbon alcohol, and dimethyl ether), and catalytic process and system integrating technologies. Large-scale industrial applications of these technologies will be gradually achieved by 2030. The core of the new catalyst researching will be focused on high efficiency control of the activation of C–C and C–H bonds and target production selectivity, which mainly relates to structure catalysis, bio-enzyme catalysis, photocatalysis, plasma catalysis, microwave catalysis, utilization of low temperature conversion, micro reactor, and other new technologies.

### 3) Ethylene producing from natural gas direct coupling reaction

Ethylene is an important organic chemical material, and also serves as an indicator of a country's development level. The shortage of petroleum resources brings in the urgent demands for the preparation of ethylene from natural gas via direct coupling reaction. It uses high efficiency catalytic technology to achieve direct conversion and utilization of natural gas, which has obvious advantages in reaction thermodynamics. The raw materials can be obtained from natural gas, hydration natural gas, and also from coal gasification methanation. The ethylene products are basic materials for existing chemical industry, and thus the industrialization of this technology will change the highly dependent of existing ethylene industry on petroleum resources completely. With 20 years of technology accumulation, the researchers have obtained great progress domestically and abroad and carried out preliminary exploration in industrialization process. Our researches in the field of catalyst are obviously superior to those of other nations. Due to the shortage of petroleum resources, this technology is expected to achieve large-scale production during 2030~2040.

### 4) Cost-effective $CO_2$ separation and recovery technology

$CO_2$ separation and recovery technology is the premise of $CO_2$ discharge reduction and green-based utilization. At present, the key of $CO_2$ separation is the cost. Therefore, new adsorbent and adsorption materials are expected to be the breakthroughs in this field. For example, molecular design of new ionic liquid adsorbent, innovation of membrane separation, mesoporous materials, and amine adsorbent bonding technologies can effectively increase adsorption capacity and help to reduce the separation cost. On the other hand, researches of regeneration method for adsorbent (or absorption) are also critical to reduce the cost of $CO_2$ separation and recovery. The realization of cost-effective adsorbent (or absorption) regeneration and $CO_2$ high concentration recovery will be attractive in future researches. It can be predicted that the $CO_2$ separation and recovery technology will achieve breakthroughs in the next 20~30 years.

### 5) High-value chemical manufacturing technologies for industry chain constructing from $CO_2$

$CO_2$ high-value and green utilization is the developing direction of $CO_2$ utilization. Clean production new technology, which starts from $CO_2$, using urea for carbonate ester synthesis, followed by further synthesizing of series of high-value chemicals such as methyl N-phenyl carbamate, diphenyl carbonate,

    *Advanced Manufacturing Technology in China: A Roadmap to 2050*

isocyanate, and polyurethane, has achieved substantial breakthroughs in atom economy reaction designing, high efficiency catalyst and special equipment, and process reinforcement. In the next 20~30 years, it is expected that new industry chain of $CO_2$ resource-based utilization will be established.

### 6) Exploitation of natural gas hydrate and $CO_2$ storage new technologies

Natural gas hydrate is a kind of clathrate compound formed by hydrocarbon gases such as $CH_4$ and water under low temperature and high pressure. In 1960s, the Soviet Union found natural gas hydrate gas field having commercial exploitation value in the permafrost, and subsequently rich natural gas hydrate deposits in marine continental shelf and the Arctic permafrost are founded. According to preliminary statistics, the global natural gas hydrate in the shallow lithosphere range of less than 2000m, the organic carbon content is nearly twice of the sum of conventional fossil fuels (coal, petroleum, and natural gas). Both $CO_2$ and $CH_4$ have the features of synthesizing hydrate with water under certain conditions, but $CO_2$ hydration conditions is milder than $CH_4$ because of the differences of gas molecules' size and their interaction with water molecules. Therefore, $CO_2$ storage can be integrated with exploration of natural gas hydrate, namely $CO_2$ is injected into natural gas hydrate bearing strata to replace natural gas and form $CO_2$ hydrate using the difference between $CO_2$ and natural gas hydrate in bearing strata. $CO_2$ replacement method is not only a kind of possible exploitation method of natural gas hydrate, but also a kind of storage method which buries $CO_2$ in the form of hydrate. This technology also maintains the geological structural stability of original bearing strata. It is expected to achieve breakthroughs during 2040~2050.

### 7) $CO_2$ purification and carbon inverse manufacturing technologies

At present, fossil fuels are the largest sources for $CO_2$ discharge. Due to the fact that human being cannot stop using fossil fuels in the near future, scientists will research and develop new technologies to separate and capture very low concentration of $CO_2$ in air. Meanwhile, driven by solar energy and nuclear power, $CO_2$ will be decomposed to produce synthetic gas and further to produce energy sources and chemicals. If such technology can be achieved, an actual $CO_2$ balance between natural world and human society will be obtained.

## 4. Developing Roadmap

With the continuous emergence of new technologies and improvement of technical maturity, it is predicted that the rate of utilizing low carbon resources will be increased gradually and resource-based utilization technology for $CO_2$ will achieve great progress in the next 20~50 years. By 2020, the integrated and large-scale synthesis gas preparation technologies will be realized. The large-scale methanol producing technology and cascade utilization technology of coal resources will be formed. By 2030, $CO_2$ cost-effective separation, large-scale fixation, and resource-based utilization will be achieved. By 2050, coupling and cascade utilization technology for low carbon resources will be formed. The large-scale production of basic alkene materials through non-petroleum route and the greenization and reclamation utilization of high $CO_2$ industries will alleviate environmental problems caused by $CO_2$.

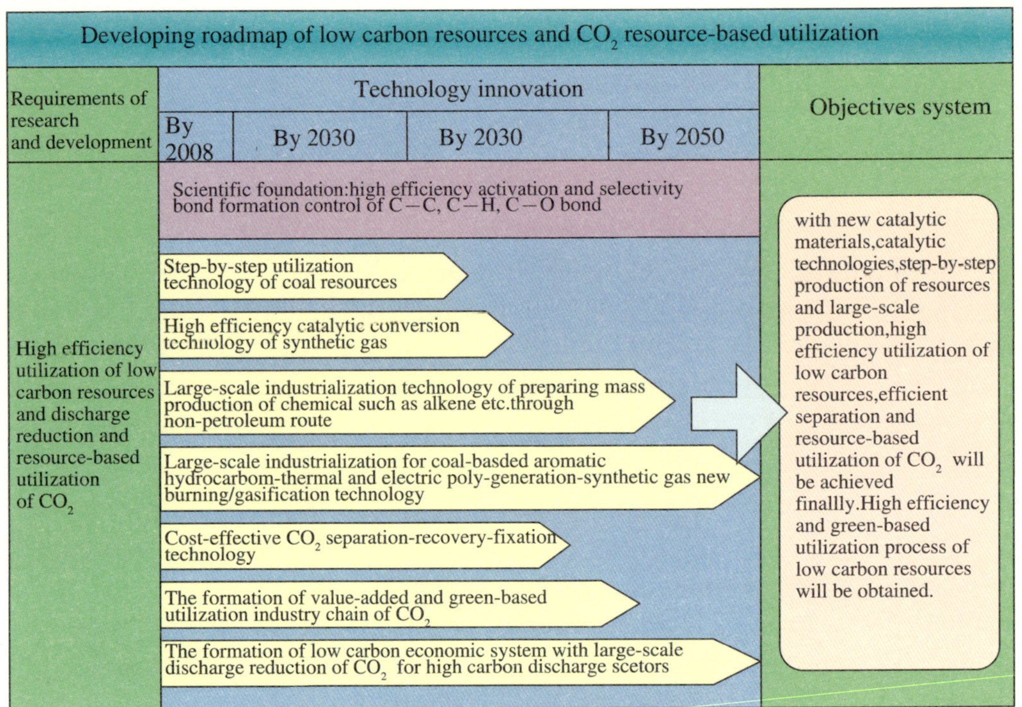



## 5. Summary

In the next 30~50 years, the breakthrough of key technologies regarding low carbon resources and large-scale application of well developed technologies can effectively promote the utilization low carbon resources. With the spreading of cost-effective $CO_2$ separation and the integrated greenization utilization technologies, internal recycling of low carbon resources will be achieved, and low carbon economic system will be preliminary established in the field of energy resources utilization.

# 3.8 Green Manufacturing Technology of Discrete Manufacturing Industry

## 1. Demands and Challenges

China is a big country of discrete manufacturing industry, which mainly includes automobile, machinery, electronics, electrical appliances industries, and etc.. The discrete industry processing technology in China is relatively laggard, and the price advantage is achieved at the cost of intensive labor, resources wasting, and environment pollution. The discrete manufacturing in China is now facing severe situation due to resource wasting, environment influence, and international green trade barriers. Thus it's urgent for us to develop new green manufacturing technology to improve the green performance, to reduce the consumption of resources, to achieve energy saving and reduction of discharge, and to break the international green trade barriers. Meanwhile, many mechanical and electrical products such as automobiles, home appliances, and machine tools, are coming into the stage of functionally scraping or technically washing out, calling for green recycling, reusing, remanufacturing, and regenerating. At present, electronic products recycling and reusing have already been developed in some areas of China. However, the recycle technologies are rudimentary, and the recovery of resources is accompanied by serious wasting and the severe secondary pollution. It's therefore urgent for us to provide support for advanced green recycling and remanufacturing technology. In developed countries, governments, research institutes, and industries all emphasize the importance of research, development, and application of green manufacturing technology in discrete manufacturing industry. For example, surface engineering technology, net shaping technology, dry processing technology, rapid prototyping manufacturing technology, and agile manufacturing technology are already in the developing stage.

## 2. Objectives and Tasks

Discrete manufacturing products are assembles of many parts via a series of discrete processes, such as mechanical and electronic equipment manufacturing  in industrial product field and electromechanical consumer products manufacturing in consumer product field. The overall goal is to realize comprehensive utilization of resources and environmental protection through breakthroughs of the key technologies. An important objective is to slow the depletion of energies and resources and achieve their sustainable utilization through integrated resource utilization, shortage resources substitution, renewable resources utilization, secondary energy utilization, and any other saving measures. Further, it is expected to decrease the generation and emissions of wastes and pollutants, improve the compatibility degree of the environment and industrial process, reduce the risk to human and environment due to production activity, and finally realize the optimal benefits of economy and environment. The immediate specific objectives are as follows:  promoting the concept of green design, realizing near 100% atom economy efficiency and selectivity in green machining process and key section in resource recycling process,  reducing the emissions of solid waste by 50% and material loss by 90%, and achieving energy-saving by 60% and material saving by over 70%.

## 3. Key Technologies

Technological breakthroughs are expected in the discrete manufacturing industry in the following aspects:

**1)Green design and assessment system**

The green design is a method of system design using advanced parallel design theory, with the help of the relevant information of products in the product life cycle, such as technical information, environmental coordination information, and economic information. And the products designed by this method have the advantages of technologically advanced, environmentally harmonic, and economically profitable.

The green design technology includes: the green product design criteria, the modular design which is based on the green environmental characteristics, the balance and long life design, the demountable design, the recovery thinking design, the design of easy maintenance and updating module and interface, the tool set of green product design, the integration of the green design tools and other design tools, and etc.. Though in the starting stage in China, the research about green design has been carried on for many years in developed countries, and green products will dominate the market in the world, with the help of gradually strengthened consciousness of environmental protection and ever growing consensus of the green consumption. Thus, it's urgent for us to expedite our pace of research and establish an evaluation system and evaluation method of green products that in accordance with our technology system, resources, environmental level, as well as industry characteristics. The research of green product evaluation mainly includes the description of green product and the research of modeling technology; the research of green products' evaluation system and method; the assessment of product life cycle and relevant evaluation standard; the product life cycle environment databases; economic databases and technologies databases; and the certification network of green product evaluation. The research about green products' evaluation system and the evaluation method has more than 20 years accumulation abroad, but little in China.

### 2) Advanced surface engineering technology

Surface engineering is a system engineering, which changes the material surface morphology, chemical composition, structure, and stress condition through coating and modifying. To develop the advanced surface engineering system, many new technologies are needed as support, such as new vapor deposition technology and equipment, nano multivariate composite coating, surface coating materials plus design process and quality control, advanced coatings technology, and surface engineering database.

For the next stage, according to the development of surface engineering technology and combining with the current needs of mechanical industry, advanced and applicable key technologies of surface engineering needed to be developed to save material substantially.

### 3) Near-net-shape forming and dry processing technology

Near-net-shape forming is a forming technology, which mechanical components only need a few or no processing after forming parts are processed by this technology. It is an integrating of new technology, new equipment, new material, and new technical achievements. And it can greatly reduce energy consumption, materials wasting, shortening the production period and reducing the cost. Dry processing is a processing mode which does not use any cooling fluid in machining process, and it is mainly used in processing machinery processing, such as dry cutting, grinding, and etc.. Due to dry processing can simplify the process, reduce cost, and eliminate the wastewater discharging and recycling, it has been successfully applied in foreign countries. On the other hand, its use in China started only in recent years, and it is expected that dry processing technology will be applied widely in the next 20~30 years.

### 4) The resource conservation and remanufacturing technology of large equipment manufacturing process

After 40 years development, the production of Chinese engineering machinery industry is only behind the USA and Japan in the world. Yet there is still a large gap compared with foreign countries, mainly reflected by low reliability of equipment, large consumption of resources, and etc.. Therefore, the research of resources saving technology in mechanical engineering is of great significance. Mainly includes:

(1) Structural light weighting design and manufacturing technology of construction machinery. Such as high performance aluminum alloy, alloy Al-Li alloy, high-strengthen magnesium alloy, titanium alloy, high temperature alloy, and composite material processing, forming, and connecting technology; the casting, forging, and fast prototyping technologies of large, complex, and overall structure integrated manufacturing technology; and etc.

(2) The reliability design of engineering mechanical system.

(3) The security technology of engineering mechanism. Such as engineering mechanical assessment and evaluation system, the research of residual life assessment system, the maintainability design of mechanical engineering, proactive maintenance, and repair design. According to research in advanced surface, the net-shape forming technology, and quality control, combining with the failure analysis, detection , diagnosis, and evaluation methods, the remanufacturing engineering of engineering machinery and electrical products improve the technical ability of remanufacturing for the overall machine, engine, transmission device, and other heavy equipment components. The resource saving and remanufacturing technology will play an important role in our nation's sustainable development of engineering machinery, and this technology is expected to be used widely in the next 10~20 years.

### 5) The bionic green manufacturing and intelligent self-repairing technology

Life science will have a significant impact on the progress of technology and development of society in the 21st century. The fusion of mechanical science and life science will generate new technologies and products, and form a new industry with great development potential.  The main contents of this research include: the engineering manufacture of biological tissues, the bionic design technology, and bionic manufacturing systems; the bionic micro mechanical and biological manufacturing process, the advanced technology from the restoration with adaptive, diagnosis, and healing; the structure design of mechanical system with the function of self-repairing; and the self-repairing additives of nano dynamic coefficient. Developing the research of bionic green manufacturing and intelligent self-repairing technology is very significant to meet the needs of national security, aerospace, and other domestic industry departments.

### 6) The dismantlability, recycling, and remanufacturing technology of waste electronic equipments

Discarded electronic appliances and the electronic wastes of remanufacturing consist mainly of discarded appliances, communication tools, battery products, and etc., covering almost all aspects of life. The electronic wastes is one of the fastest growing solid wastes in recent decades, and the recycling of which is beneficial for the maximum utilization of waste resources. To solve the problem of dismantlability, recycling, and reduction, two aspects should be explored. One is to utilize the green design technology, exploring the energy-saving platform and modeling design technology, the green design technology for dismantling and recycling, the easy maintaining and upgrading module and interface design technology, and etc.. The other one is to research the recycling technology of waste electronics, including disassembly technology, materials circulation technology, and etc..

### 7) The dismantlability recycling and remanufacturing technology of automobile

The dismantlability recycling and remanufacturing technology of automobile, which fulfills components reuse and materials recycling in an economical way when automotive products reach their life-cycle, is an important material technology to realize green manufacturing. Automobile's green remanufacture engineering is a way to achieve regeneration of high quality products and full utilization of waste automotives by applying advanced surface technology and other cutting edge repairing and transforming technologies, as well as strict product quality management and market management model. For the automotive industry, our study focuses on the equipment for parts disassembling, typical parts cascade utilization and green remanufacturing technology, technology and devices for recycling and fast green separation of the automobile's aluminum alloy, technology for recycling the discarded automobile's non-metallic materials, clean technology for recycling platinum group metals from automobile's waste catalyst, and facelift technology and equipment for injection molding tires. On the basis of automobile enterprise's self-production policy, large-scale demonstration project for vehicle parts remanufacturing is expected. Through remanufacturing, the overall quality and performance of automotive remanufacturing products will meet or exceed the new standard; in comparison with the production of new

   *Advanced Manufacturing Technology in China: A Roadmap to 2050*

parts, the cost will decrease 50%, energy consuming will decrease more than 60%, and materials expenditure will decrease more than 70%. It is the second investment in automotive products and important steps for automotive products resource reusing and appreciation. The key technologies include: the depth of disassembling and sequence control technology; automatic control technology for disassembling; online identification and classification technology; evaluation of product remanufacturing design and remanufacture feasibility; the analysis technology for structural failure of remanufacturing materials and structure; evaluating methods of remaining obsolete equipments and prediction method of remanufactured components' life; nano-surface engineering technology; quality control technology for remanufacturing; integration technology for advanced materials' manufacturing and forming; technology for virtual remanufacturing; technology for advanced non-destructive testing and evaluating; and technology for fast prototyping in remanufacturing.

### 8) The green packaging technology and green materials

The green packaging technology means to optimize packaging process with full consideration of environmental protection and economic feasibility by choosing environmental friendly packaging materials which can be reused, recycled, decomposed, and cause no harm to human being and the environment during the whole life of the products. The goal of such technology is to minimize the consumption of resource and the generation of waste. The green packaging technology includes the design of green packaging, the choices of material, the recycling and disposal, legal regulation and environmental sign, and etc. Green material is ecological environment material, and is the material with good property and function, therefore beneficial to human health. It consumes less resources and energy, causes less pollution to ecology and environment, and can coexistence with the environment in its whole life cycle. Green material is the important base of green manufacturing, and it is also the material base and the core of green packaging technology. The research of green packaging technology and green materials has been developed for decades in the developed countries, but is in the initial stage in our country. Developing green packaging is the trends of packaging industry, and accelerating the industry system of green packaging is an inevitable way for our packing industry. The key technologies of green packaging include the advanced production technology of green packaging and green materials, the technology of degradation, recycling and reuse, the technology of substituting, the treatment and comprehensive utilization technology of packaging wastes, and so on. The green packaging technology and green materials will be widely used in the next 20 years.

### 9) Green supply chain and green logistics

Supply chain is critical in international competition globally; and accompanied with the ever increasing public reorganization of environmental issues and restriction of legislation, environmental problems have become vital in developing supply chain. For instance, after the publication of the EU Rohs order and WEEE order, the global electronic industry has experienced dramatic changes. Due to years of accumulation of technologies and policies in the developed countries, many policies such as the producer responsibility policy in the EU and the consumer responsibility policy in Japan, are already in operation in their areas. On the basis of the experiences from developed countries and in conjunction with our unique national condition, it is expected to establish an integrated production system combing 'design →manufacture→using→ discard' arterial system and a 'purchase→decomposition→choose → reuse→manufacture' venous system, realizing the internal circle from forward supply chain to reverse supply chain. Establish legislations to clarify the producer responsibility in handling and recycling wastes; deeply plant the philosophy of 'design →manufacture→using→ discard' into every section of the production industries, achieving 90% recycleing of materials. It is expected that in the next 20~30 years, green logistics will be widely applied in global manufacturing industries.

### 10) Establish sustainable manufacturing system with combination of arteries and veins

Arterial system is a production system with its process: 'design →manufacture→using→ discard'; venous system is a resource recycle system with its process: 'purchase→decomposition→choose → reuse→manufacture'. The goal of establishing this system is to formulate policies, laws, regulations, and standards in accordance with the situation of China, and to establish sustainable manufacturing system with combination of arteries and veins. The main contents of this research include: policies, laws and regulations, and standard system for product using, recycling, and remanufacturing; the responsibility system related to the products' whole life cycle; the management style of scrap recycling; qualification and certification system and quality standards for recycling and remanufacturing enterprises; and the legislations for product recycling, processing responsibility.

At present, developed European countries and Japan have established a set of corresponding standard systems, but the systems are not perfect, still existing some problems have not been solved. For example, the sustainable manufacturing system in some industries is still unachievable. There is still no effective solution to decrease the cost that resource recycling brought to enterprises. The corresponding regulations system laggard, consequently, enterprises lose the advantages in the international competitive. Our current manufacturing resource consumption is higher than that of foreign countries by 40%, and hence it's significant for us to realize the target of resource-conserving manufacturing. It is also the important foundation for building the resource saving society and realizing the sustainable development.

## 4. Developing

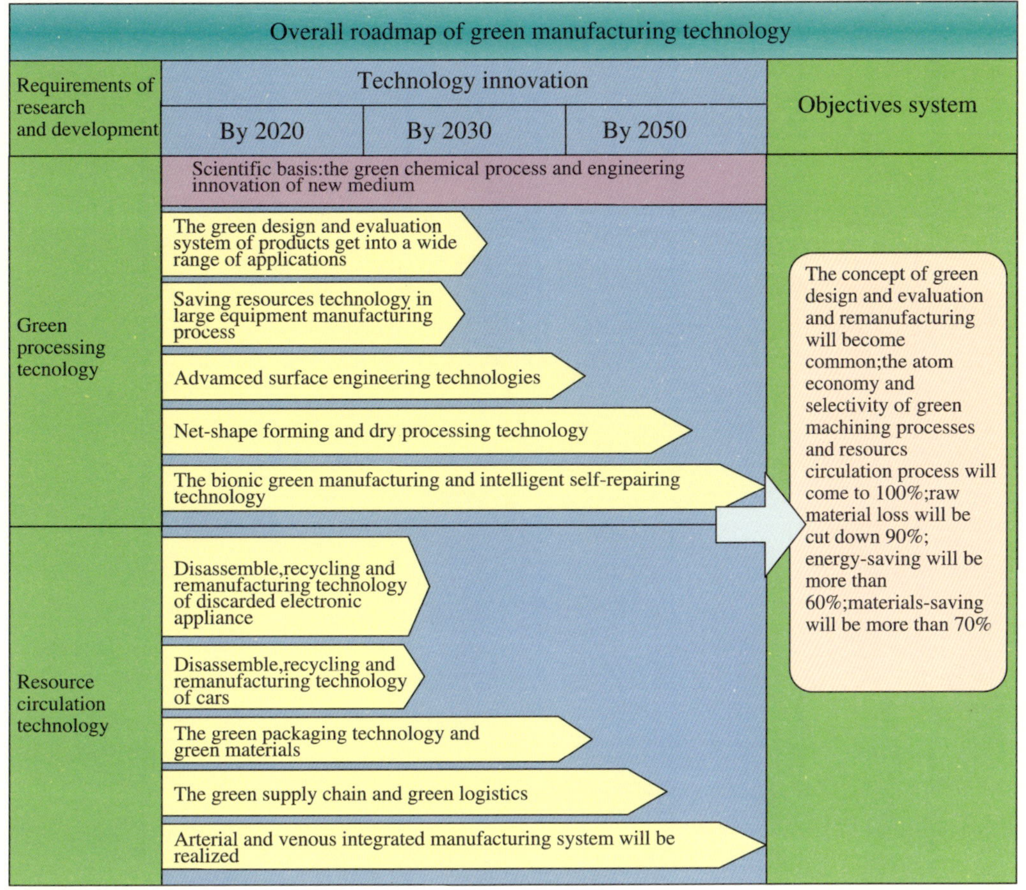



## 5. Summary

In a word, it is estimated that the green product design and evaluation system will be widely used by 2020. And the green product evaluation system and evaluation methods that match our technologies, resources, and environment level as well as industry characteristics will also be established by 2020. At that time, the evaluation of green products will be started. And by 2030, we will achieve many technologies for large equipment manufacturing process, such as resource saving technology, advanced surface engineering technology, electronic electric appliance, the wasted car disassembling and reclamation, and the green packaging technology. By 2050, near-net-shape forming and dry processing technology, the bionic green manufacturing, and intelligent technology will be formed; the integration of arterial/vein production system will be realized. Through allocating resources properly, maximizing the manufacturing system efficiency, using different methods, we will finally realize the green manufacturing target of resource conservation of energy and environmental protection.

Discrete manufacturing: According to the characteristics of product manufacturing process, manufacturing can be summarized as continuous manufacturing and discrete manufacturing. Compared to the products of continuous manufacturing, discrete manufacturing products are always assembled by a series of components that processed after a series of incontinuous procedures. The enterprise that process and sale such products can be called as discrete manufacturing enterprise. For example, both mechanical and electronic equipment manufacturing which belong to the production of material production and the electromechanical products manufacturing which belongs to the production of the consumption goods are discrete manufacturing.

Discrete Manufacturing firms typically contains processes such as parts processing, assembling components into products. In these enterprises, some emphasis on the parts and components manufacturing and we call them discrete processing enterprises; Some emphasis on the assembly and we call them assembly-oriented enterprises; In some enterprises, the processing and assembly are completed mainly on their own, for such enterprises, regardless of their size, they are more complex than the previous two types of enterprises. Discrete manufacturing industry is industrial chain network. which consist of interrelated supporting of thus three categories of enterprises.

Production processes in discrete manufacturing are usually separated into a lot of processing tasks. Each task only requires a small part of capacity and resources. Enterprises will generally established organizations (departments, workshop section, or groups)by making use of some functionally similar equipments in accordance with space and administrative management. In each department, work pieces are processed in different types of processes from one work center to another. Enterprises often arrange the location of equipments in accordance with the main process in order to minimize the materials' transmission distance. In addition, the processing technology and equipments utilization are also very flexible, product design, processing requirements and quantity of orders often change.

Remanufacturing: remanufacturing is short for remanufacturing engineering. Remanufacturing is a series of technical measures and engineering activities that repair and transform waste mechanical and electrical products, guided by electromechanical products' full lifecycle design and management, achieving the goal that increases mechanical and electrical products' performance tremendously, obeying the standard that high quality, high efficiency, energy saving, and environmental protecting, by means of advanced technology and industrialization.

The important feature of remanufacturing is the quality and performance of manufactured products can meet or exceed the new products, cost is about 50% of new products, energy saving is more than 60%, materials saving is more than70% and a significant contribution to the environmental protection; remanufactured products are not second-hand products but new products. Manufacturing engineering has become an important part of developing the circular economy and building an economized society, under the guidance of national sustainable development strategies and the scientific concept of development that 'people oriented, comprehensive, coordinated, and sustainable development of population, resources, and environmental'.

# References


[1] Bai E Z, Hu Y G, Lin Y H, et al. 2005. Research progress of C1 chemical technology. Petrochemical Technology, 34(Z1): 34-37.

[2] Cai Q R, Peng S Y, et al. 1995. Catalytic effect in C1 chemistry. Beijing: Chemical Industry Press.

[3] Chen H Z. 2008. Biomass Science and Engineering. Beijing: Chemical Industry Press.

[4] Chen H Z, Wang L. 2008. Research progress on key process and integrated eco-industrial chains of biobased products —— Proposal of biobased product process engineering. The Chinese Journal of Process Engineering, 4: 676-681.

[5] Chi X B, Zhao Y. 2007. High performance computing technology and its applications. Bulletin of The Chinese Academy of Sciences, 22(4), 306-313.

[6] Cui Y, Ma X N, Lv H. 2008. Perspective of manufacturing industry on the era of globalization (Part 1). People's Daily•International weekly (6th Edition), 2008-4-25.

[7] National Development and Reform Commission, Ministry of Finance, State Administration of Taxation. 2004. Resources comprehensive utilization catalogue (Revised in 2003). Development and reform environmental resources [2004]73.

[8] National natural science foundation of China-Engineering and material science department. 2006. Metallurgy and Material Preparation Engineering Science. Beijing: Science Press.

[9] Huang K, Jiang S, Ma T C, et al. 2008. Developing strategy of key technologies field of British high value manufacturing. National Science Library, Chinese Academy of Sciences, Bulletin of dynamic monitoring of scientific research, 14(60): 1-8 .

[10] Li J H, Guo M S. 1999. Quantitative science approaches of process engineering —— multi-scale method. Progress in Natural Science, 9(12): 1073.

[11] Li Y F, Dai Y N, Liu H X. 2007. Recovery and utilization of nonferrous metal's secondary resource in China. Mining and Metallurgy, 16 (1): 86-89.

[12] Liu F. 2005. Theory and Technology of Green Manufacturing. Beijing: Science Press.

[13] Liu G F. 2001. Green Design and Green Manufacturing. Beijing: China Machine Press.

[14] Lu Y X. 2005. Challenges and opportunities Chinese manufacturing industry faces in 21st century. Mechanical Engineering, 1: 9-13.

[15] Mao Z S, Chen J Y. 2004. Engineering Foundation of Chemical Reaction. Beijing: Science Press.

[16] Qu Y X. 2006. Numerical Simulation and Software of Chemical Process. Beijing: Chemical Industry Press.

[17] Wang B S. Translation of specialized technical terms of secondary resources. China Terminology, 10(4): 51-54.

[18] Wang F S, Wang K, Chen X Z, et al. 2000. Prospect of Chemistry in 21st Century. Beijing: Chemical Industry Press.

[19] Wu C Z, Ma L L. 2003. Modern Utilization Technology of Biomass Energy. Beijing: Chemical Industry Press.

[20] Xu B S. 2005. Remanufacturing Engineering Basis and Its Application. Harbin: Harbin Institute of Technology Press.

[21] Xu B S. 2007. Remanufacturing and Cycle Economy. Beijing: Science Press.

[22] Xu K D. 2005. The current state and challenges of Chinese manufacturing industry. China Development Observation, 4: 13-15.

[23] Xu S S, Zhang D L, Ren Y Q. 2006. Large-scale coal gasification technology. Beijing: Chemical Industry Press.

[24] Zhang C. 2004. Visit research group leader of ecological construction, environmental protection and recycling economy science and technology issues-Honglie Sun. http://www.stdaily.com/gb/

[25] Zhongchang/2004-07/06/content_271396.htm.

[26]　Zhang Y. 2005. The green process engineering science. The Chinese Journal of Process Engineering, 1(1):10-15.

[27]　Zhang Y, Xu B S, Duan G H. 2006. Focus and trend of green manufacturing technology development. High Technology Development Report of Chinese Academy of Sciences. Beijing: Science Press.

[28]　Zhang Z S, Cui G X, Xu C X. 2005. Turbulence Theory and Simulation. Beijing: Tsinghua University Press.

[29]　China nonferrous metals industry association. 2007. Progress in Nonferrous Metals. Changsha: Central South University Press.

[30]　China Society and Technology Association. 2007. Developing Report of Metallurgy Engineering Technology Discipline. Beijing: China Science and Technology Press.

[31]　Chinese Business Newspaper Network. 2008. Development strategy research report of China nonferrous metals enterprise.

[32]　Zhong S Q, Xiong J Y, Zhang Y Z, et al. 2005. Strategic research on unconventional energy source of china in 21st century. Drilling & Production Technology, 28(5): 93-98.

[33]　Zhu Q S, Yan L F, Guo Q X. 2002. Biomass Clean Energy.Beijing: Chemical Industry Press.

[34]　Agrawal R, Singh N R, Ribeiro F H, et al. 2007. Sustainable fuel for the transportation sector. Proceedings of the National Academy of Sciences of the United States of America, 104: 4828-4833.

[35]　Amenomiya Y, Birss V I, Goledzinowski M, et al. 1990. Conversion of methane by oxidative coupling. Catal Rew Sci Eng, 32(3): 163-227.

[36]　American Chemical Society, et al. 1996. Technology Vision 2020: U.S. Chemical Industry. http://www.acs.org.

[37]　Anastas P T, Heine L G,Willamson T C. 2001. Green Engineering. Washington D C: American Chemical Society.

[38]　Dong K, Zhang S, Wang D, et al. 2006. Hydrogen bonds in imidazolium ionic liquids. J Phys Chem A, 110: 9775-9782.

[39]　Gao J J, Li H Q, Zhang Y F, et al. 2007. A non-phosgene route for synthesis of methyl N-phenyl carbamate derived from $CO_2$ under mild conditions. Green Chemistry, 9(6): 572-576.

[40]　Giaconia A, de Falco M, et al. 2008. Solar steam reforming of natural gas for hydrogen production using molten salt heat carriers. AIChE J, 54(7): 1932-1944.

[41]　Grossmann E, Westerberg A W. 2000. Research challenges in process systems engineering.AIChE J, 46(9): 1700.

[42]　IChemE. 2007. A Roadmap for 21st Century Chemical Engineering. www.icheme.org/TechnicalRoadmap.

[43]　Klaus S L. 2003. A guide to $CO_2$ sequestration. Science,300: 1677-1678.

[44]　Klemes J, Huisingh D. 2005. Make progress toward sustainability by using cleaner production technologies, improved design and economically sound operation of production facilities. Journal of Cleaner Production, 13(5): 451-454.

[45]　Li J H, Kwauk M. 2003. Exploring complex systems in chemical engineering—the multi-scale methodology. Chemical Engineering Science, 58(3-6): 521-535.

[46]　Metz B, Davidson O, de Coninck H, et al. 2005. IPCC Special Report on Carbon Dioxide Capture and Storage. Cambridge: Cambridge University Press.

[47]　Mineral Processing Technology Roadmap. 2000. http://www1.eere.energy.gov /industry/mining/pdfs/mptroadmap.pdf.

[48]　New Process Chemistry Technology Roadmap. 2001. http://www.Chemicalvision 2020. org /pdfs/new_chemistry_roadmap.pdf.

[49]　Roberts C B, Elbashir N O. 2002. Advances in C1 Chemistry in the Year 2002. Fuel Processing Technology, 83.

[50]　Sarkis J. 2001.Greener Manufacturing and Operations: From Design to Delivery and Back. UK:

Greenleaf Publishing.

[51] Taniewski M. 2004. The challenges and recent advances in C1 chemistry and technology. Polish Journal of Applied Chemistry, 48: 1-21.

[52] Thompson T B, Kontomaris K. 1999. Chemical Industry of the Future: Technology Roadmap for Computational Fluid Dynamics. U.S. Department of Energy Office of Industrial Technology.

[53] Zhang X, Li C, Fu C, et al. 2008. Environmental impact assessment of chemical process using the green degree method. Ind Eng Chem Res, 47: 1085-1094.

[54] Zhang Y H, Yang C, Mao Z S. 2008. Large eddy simulation of the gas-liquid flow in a stirred tank. AIChE J, 54(8): 1963-1974.

[55] Zoebelein H, Böllert V. 2000.Dictionary of Renewable Resources. Weinheim: Wiley-VCH Verlag GmbH.